D0023259

HOW THE LASER HAPPENED

HOW THE LASER HAPPENED

ADVENTURES OF A SCIENTIST

Charles H. Townes

New York Oxford

Oxford University Press

1999

Oxford University Press

Oxford New York
Athens Auckland Bangkok Bogotá Buenos Aires Calcutta
Cape Town Chennai Dar es Salaam Delhi Florence Hong Kong Istanbul
Karachi Kuala Lumpur Madrid Melbourne Mexico City Mumbai
Nairobi Paris São Paulo Singapore Taipei Tokyo Toronto Warsaw

and associated companies in
Berlin Ibadan

Published by Oxford University Press, Inc.
198 Madison Avenue, New York, New York 10016

Library of Congress Cataloging-in-Publication Data
Townes, Charles H.
How the laser happened : adventures of a scientist / by Charles H. Townes.
p. cm.
Includes index.
ISBN 0-19-512268-2
1. Lasers—History. 2. Masers—History. 3. Science and state—
United States—History. I. Title. II. Title: How the laser happened.
QC887.2.T68 1999
621.36'8'09—DC21 98-22216

3 5 7 9 8 6 4 2

Printed in the United States of America
on acid-free paper

INTRODUCTION AND ACKNOWLEDGMENTS

The Alfred P. Sloan Foundation initiated this book with a request that I write a history of the laser. That history is here and with it is also much of my own personal story as a scientist. The latter is because the modern development of science and technology is intimately dependent on people, their interactions, and mutual stimulation—what can be called scientific sociology. Something like the laser development does not come from an isolated idea, but from a scientific milieu, curiosity, struggles and puzzles, and from many different people.

Interactions with government and military policy and with international affairs are also part of the laser and the science story; they are areas in which scientists have been increasingly involved during the twentieth century.

My rather intense work in science and at times public service, including a number of changes from one job and location to another, have been generously and thoughtfully supported by my wife Frances, to whom I and the story reported here are very much indebted. This book also owes much to the science writer Charles Petit, who has helped me enormously in organizing and drafting it and in giving sensitive editorial advice. Without him, it could not have been done. And I thank my capable secretary, Marnie McElhiney, for much help in the book's preparation.

My hope is that this account will give a realistic and interesting illustration of the way that ideas and science may really come about and of a scientific career in the present age.

Charles H. Townes
Berkeley, California

CONTENTS

HOW THE LASER HAPPENED

1

THE LIGHT THAT SHINES STRAIGHT

On July 21, 1969, astronauts Neil Armstrong and Edwin Aldrin set up an array of small reflectors on the moon and faced them toward Earth. At the same time, two teams of astrophysicists on Earth—240,000 miles away—at the University of California's Lick Observatory and at the University of Texas's McDonald Observatory, prepared small instruments on two big telescopes. They took careful note of the location of that first manned landing on the moon. About ten days later, the Lick team pointed the telescope at that precise location and sent a small pulse of power into the tiny piece of hardware they had added to the telescope. A few days later, after the west Texas skies had cleared, the McDonald team went through the same steps. In the heart of each telescope, a narrow beam of extraordinarily pure red light emerged from a crystal of synthetic ruby, pierced the sky, and entered the near vacuum of space. The rays were still only about 1,000 yards wide after traveling the 240,000 miles to illuminate the astronauts' reflectors. Slightly more than a second after light hit the reflectors, the crews in California and in Texas each detected the faint reflection of its beam. The interval between launch of the pulse of light and its return permitted calculation of the distance to the moon within an inch, a measurement of unprecedented precision.

The ruby for each source of light was the heart of a laser, a type of device first demonstrated in 1960, just nine years earlier. Even before man reached the moon, an unmanned spacecraft had landed on the moon in January, 1968, with a television camera that detected a laser beam shot from near Los Angeles by the California Institute of Technology's Jet Propulsion Laboratory. That beam radiated only about one watt. But from the moon, all the other lights in the Los Angeles basin, drawing thousands of megawatts, were not bright enough to be seen. Their light spread and diffused into relative indetectability while that single beam, with the power of a pocket penlight, sent a twinkling signal to the lunar surface.

Laser beams reflected from the moon, allowing measurement of the moon's distance, is only one illustration of the spectacular quality of laser light. There are many others, as well as myriad everyday uses for the laser. But for several years after the laser's invention, colleagues used to tease me about it, saying, "That's a great idea, but it's a solution looking for a problem." The truth is, none of us who worked on the first lasers imagined how many uses there might eventually be. This illustrates a vital point that cannot be over stressed. Many of today's practical technologies result from basic science done years to decades before. The people involved, motivated mainly by curiosity, often have little idea as to where their research will lead. Our ability to forecast the practical payoffs from fundamental exploration of the nature of things (and, similarly, to know which of today's research avenues are technological dead ends) is poor. This springs from a simple truth: new ideas discovered in the process of research are really *new*.

As soon as we had masers and lasers, I argued that they marry two very important and widely used fields of science and technology: optics and electronics. While we could not predict all the different places these devices would take us, we could hence expect that they would have a wide range of applications, which is exactly what happened.

Once invented, lasers found myriad uses. The device that shot the moon's distance was a middle-sized laser, and by the time it pulled off that feat surveyors were already using lasers for such mundane, but perhaps more useful, tasks as laying out land boundaries or grading roads. The Bay Area Rapid Transit (BART) trains that cross under San Francisco Bay do so in an underwater tube along a laser-laid path. I know exactly where the borders of my farm in New Hampshire are, and their precise length, because the surveyor used a laser.

The smallest lasers are so tiny one cannot see them without a microscope—thousands can be built on semiconductor chips like those that form the hearts of computers (before long, some computers may in fact use light from lasers in the way that computers now use electrical impulses). The biggest lasers consume as much electricity as a small town. About 45 miles from my office at the University of California in Berkeley, is the Lawrence Livermore National Laboratory, which has some of the world's presently most powerful lasers. One set of lasers, collectively called NOVA, is set up so that ten individual laser beams converge in a spot the size of a pinhead. The lasers themselves are immense things that stretch for more than 400 feet in a train of powerful electrical coils, optics, and thick lavender glass plates 20 inches in diameter, from which the laser bursts arise. As they converge, their focused beams of light almost instantly (literally, in a few billionths of a second) create temperatures of many millions of degrees. Such intensely concentrated energies are essential for experiments that could show physicists how to create conditions for nuclear fusion, the process that makes the sun shine. The Livermore team hopes thereby to find a way to generate electricity efficiently, with little pollution or radioactive waste. The team is also progressing toward a still more powerful laser, the National Ignition Facility (NIF), and with their present laser have just increased the world's record for power by a factor of 10, reaching a million billion watts, again with beams focused on a tiny speck. This pulse lasts a little less than a trillionth of a second, but while it lasts its power is enormously greater than the power consumed at that moment by our entire globe.

While the laser fusion program at Livermore may be an exotic example of the way laser beams can alter materials, there are hundreds of more mundane materials processing applications. For example, the bearing surfaces in your car may have been treated by laser beams played across them—a laser beam heats up steel so fast that it hardens the surface without appreciably heating the interior, which would make the bearing brittle. Lasers can evaporate and thus remove material so quickly that any neighboring material is not affected at all by heat. The laser's focused intensity easily penetrates diamond, our hardest material. Lasers are precise, too, cutting tiny holes in rubies used as bearings in fine Swiss watches and refining the intricate patterns of electronic circuitry in computer microchips. Their fantastically short pulses can cut and evaporate material so quickly that the remaining material is undisturbed.

Contrasting with the power and intensity that laser beams have achieved, scientists have also found that a weak laser beam focused by a microscope can gently move tiny particles around, including even the organelles inside living cells. Such "optical tweezers" can be powerful tools for biological research. Lasers are also used to slow the high-speed motion of atoms and hold them in traps, creating pockets of gas with temperatures of only a few billionths of a degree above absolute zero, the lowest temperatures yet achieved by researchers.

Agencies that monitor air pollution can instantly learn, in the field, how dirty the air is over a city-wide basin, by comparing the ways laser beams of various colors are absorbed. They can check the air over smokestacks to determine the pollutants coming from them. They can even look straight up to measure certain chemicals in the stratosphere.

Lasers provide an unsurpassed medium for communication. One laser beam can, in principle, carry all the information that is passing back and forth right now among all the people and computers in the world. The signals from all the telephone lines, all the television stations, all the radio stations, all the talk and music and digitized information—all could be packed into just one laser beam. This possibility has not yet been achieved, but we are well on the way to do so and already use remarkable rates of communication by laser beams. In addition, a laser beam can be sent through a tube of flexible fiber-optic glass that is narrower than a pencil lead. The small size of fiber-optic cables, quite aside from their immense capacity, is a reason why new telephone lines under the streets of New York City rely on fiber optics. The utility conduits are already so crowded with sewers, power lines, phone lines, television cables, and other arteries vital to modern society, that it would be difficult to put in new copper transmission lines, even if one wanted to. Plus, laser communications provide greater secrecy than standard radio or telephone transmissions. A laser beam goes in a straight line, unlike a radio beam that spreads widely and is easy to intercept. A laser beam in a fiber-optic cable cannot be intercepted unless someone attaches a detector to the fiber itself.

Tens of millions of Americans have lasers in the their homes, and many have them in their cars—lasers that make music. Information on a compact disc (CD) is recorded by a laser beam, and laser beams inside CD players read from them the digitally encoded sound that makes a CD such a faithful music medium. Laser artistry has been extended to colorful light shows, with laser beams crossing in the sky and painting patterns. Objects can have their three-dimensional shapes recorded on photographic film, by laser beams reflecting off them, producing a hologram—and three-dimensional–appearing images of the objects can be reassembled, seeming to hover in space, when other laser beams are reflected off the pattern recorded on the film. Some uses of lasers are mundane, but they also illustrate this technology's versatility—not long ago, for example, Battelle Memorial Institute in Columbus, Ohio, announced that lasers make very good potato peelers.

Since their invention, lasers have provoked people to imagine them turned by the military into death rays, and a great deal of money has actually been spent trying to use lasers to destroy incoming missiles. It is not clear how practical such a system can be, but if it works, henceforth we may not need to fear missile attack, and that would be a blessing. The idea of flashing death rays also has a mystique that catches human attention; and so we have Jove's bolts of lightning and the death rays of science fiction, beginning at least as early as Alexei Tolstoy's novel of 1926, *The Garin Death Ray*:

> Garin turned the machine towards the door. On the way the ray from the apparatus cut through the electric lighting wires and the lamp

> on the ceiling went out. The dazzling, dead straight ray, as thin as a needle, played above the door and pieces of wood fell down. The ray crawled lower down. There came a short howl, as though a cat had been trodden on. Somebody stumbled in the dark. A body fell softly. The ray danced about two feet from the floor. There was an odor of burning flesh.—Garin coughed and said in a hoarse voice . . . "They're all finished with."

Such pictures of powerful and instantaneous beams clearly attract human attention, whether real or not. Although modern laser beams can seem very similar to these mystical rays of fiction, lasers don't make very handy death rays; in most cases it is much more effective and easier to hit something, or somebody, with a standard weapon (or a rock) than with a laser.

Actual military uses of lasers as death rays are likely to be rare and specialized, if they occur at all. Yet there are indeed important military uses for lasers. Communication is one, which of course is not peculiar to the military. The ability of modern tanks, bombs, and missiles to hit their targets precisely, within a few feet, is important, and that ability often depends on lasers. A laser beam aimed at the target provides, as it reflects off the target or guides a missile, a distinct beacon on which a warhead can home. The thing that pleases me particularly about military lasers is that, if war must happen, and such a target as a bridge must be destroyed, lasers help send the bombs just to the bridge. The chances are much reduced that bombs will hit the surrounding neighborhood. Videos from the 1991 Gulf War against Iraq, showing military targets destroyed with little or no damage or loss of life in the civilian areas next door, recorded this remarkable precision.

Hospitals and some doctors' offices are full of lasers. Surgeons operate on detached retinas at the backs of eyes by sending laser beams through patients' pupils. The laser light has little effect until it is absorbed by the pigments of the retina, thereby creating a small scar and effectively welding it into place. I have a special emotional reaction when a friend tells me that a laser operation has saved his or her eyesight—very different from hearing of other purely technical laser successes. In all manner of surgery, the ability of lasers to cut tissue and stop bleeding at the same time, by cauterizing blood vessels, has made the jobs of surgeons easier and the lives of their patients more secure.

Lasers not only do practical jobs but also are powerful and versatile research tools. In scientific research, one of the laser's great strengths is its high precision. A laser produces light waves of extreme purity, having a nearly uniform wavelength and extraordinary directivity. The light waves are coherent—a technical term meaning that all the light has not only the same wavelength but that all the waves are in step with each other. This allows for a precision in the use of light and other electromagnetic radiation that scientists could not, before the laser and the maser, ever have expected.

The maser, which produces microwaves (the shortest variety of radio waves), initiated the basic idea behind lasers. The laser produces shorter wavelengths of electromagnetic radiation—infrared radiation and optical light—rather than microwaves. Lasers have therefore overshadowed the earlier masers. However, masers are our most sensitive amplifiers of microwaves and, as atomic clocks, also provide some of our most accurate measurements of time.

Light waves are forms of electromagnetic radiation that, as the name implies, are traveling, linked electric and magnetic fields. Focused to a small point, laser beams can produce intensities of light billions of times that at the sun's surface, with a correspondingly intense electric and magnetic field at the focal point; this capability has created another entirely new field of physics and engineering called nonlinear optics. The transparency and other optical qualities of materials change in the presence of intense laser beams. Nonlinear optical effects also allow intense laser beams to pass through the atmosphere or other optical media without the spreading pattern of weaker beams. The intense laser light changes the optical nature of the material through which it passes, forming a sort of channel that helps to confine the beam or even focus it in a smaller pattern. A laser beam's light waves can affect materials so much that electrons in the material oscillate widely in response, emitting harmonics of the light. A ruby laser, emitting red light, can hence stimulate atoms to produce ultraviolet light, at an exact harmonic of red light.

One of the widest scientific uses of lasers has been the precise measurement of distances, and not just to the moon, but between two points in a laboratory or in a machine shop. Units of distance are now even defined in terms of wavelengths of light produced by standardized lasers; the "standard meter"—one made of a platinum-iridium alloy and kept in Paris in accordance with Napoleon's original plan for standard weights and measures—is no longer the world's length standard. Laboratory lasers now measure distances down to much less than a thousandth of the wavelength of light, well into and even beyond the realm of an atom's size. A new variety of microscopes, called scanning microscopes, move extremely fine needles across surfaces to measure the sizes and locations of individual atoms. With laser illumination through that fine needle, such a microscope sees details on a surface more than a factor of ten smaller than before. Experiments are now being set up to use lasers to help detect gravitational waves, subtle ripples in the fabric of space and time that may be triggered by the sudden movements of huge masses, such as the collapsing cores of stars during supernova explosions far away in this and other galaxies. Only a laser may provide the precise distance measurements needed to detect gravitational waves, about one billionth of an atomic diameter across distances of several miles.

Lasers give astronomers a simple way to keep their mirrors and other optical components perfectly aligned. A far more spectacular contribution of lasers to astronomy has been applied to the world's largest telescopes, involving a technique called adaptive optics, a way of looking through the atmosphere more clearly than ever before. Anyone who has ever looked at the stars knows that they twinkle; although romantic, astronomers do not like the effect. A star twinkles because Earth's atmosphere is not uniform and is always in motion—many times every second altering the path that the starlight takes through it, blurring the image. Astronomers like a star to appear clear and stationary.

A laser beam can be fired upward through the atmosphere along the path that a telescope is looking. As the laser light scatters back from small particles or atoms in its path, or from the light of atoms it stimulates into fluorescence, it carries information that can reveal the kinds of distortions present in the atmosphere along that path at that instant. The distortion is measured and converted to electrical signals that drive specially designed, flexible corrective mirrors, which can change their shapes to compensate for the fluctuating optical qualities of the atmosphere overhead. Already, such telescopes have provided images 20 to 30 times sharper than telescopes without such corrections. We are now starting to see images from the ground that are as clear as those astronomers had felt were possible only with telescopes in space—beyond the atmosphere—such as the Hubble Space Telescope. Ground-based telescopes larger than the Hubble, with much better inherent focusing abilities, should soon greatly surpass the Hubble's vision at optical and infrared wavelengths.

Also, I have recently learned of what seem to be realistic plans to try to control lightning bolts with lasers.

It is often difficult to say just when a development began, but perhaps one can say that physicists started on the path that led to the laser about 1945. No one in 1945, given a list of today's laser applications, would have guessed which of the research programs of the day was the one destined to do the trick. Yet in retrospect, it is a wonder that invention of the laser took so long. The reason will become clear as this story unfolds. However, the laser could have happened 30 years earlier than it did. And one can, it turns out, make a laser out of almost anything. My friend and brother-in-law Arthur Schawlow, who played a fundamental part in this whole story, wanted to illustrate this by making an edible laser. He tried Jell-O, but that didn't work, so he mixed regular gelatin with a fluorescent dye, and it worked just fine.

This stunt was paralleled by a drinkable laser made with tonic water by Eastman Kodak Company researchers. The late Richard Feynman, a superb physicist, said once as we talked about the laser that the way to tell a great idea is that, when people hear it, they say, "Gee, I could have thought of that." We all should have thought of the laser much sooner.

By now, six Nobel Prizes in physics have involved uses of masers and lasers, including one that stemmed from their invention (Nikolai Basov, Alexander Prokhorov, and Charles Townes in 1964; Dennis Gabor in 1971; Arno Penzias and Robert Wilson in 1978; Nicolaas Bloembergen and Arthur Schawlow in 1981; Norman Ramsey in 1989; and Steven Chu, Claude Cohen-Tannoudji, and William Phillips in 1997). I expect there will be more, particularly in chemistry and biology. Of course, one could in principle say the same about the usefulness of the lowly screw-driver, a tool used in many experiments; but the point is that lasers and masers have newly arrived and are now very important scientific tools. They have changed our scientific capabilities and helped scientists accomplish some spectacular feats.

The list of applications goes on and on. And while lasers are made from a variety of materials and in various forms, they share certain basic principles of physics. The chapters to come tell the story of how the laser and its predecessor, the maser, came to be. Right here is perhaps the best place to describe in plain language just what these devices are.

How Lasers Work

LASER is an acronym for light amplification by stimulated emission of radiation. Lasers (and masers, an acronym for microwave amplification by stimulated emission of radiation) rely on fundamental ways that radiation interacts with molecules, atoms, or electrons. We all are familiar with what usually happens when light shines on something: part of the light gets absorbed. If one shines a bright light on a piece of black paper, its molecules absorb the light, receive its energy, and heat up. A laser turns this process around. It is as though the black paper, instead of getting energy *from* the light, gives energy *to* the light. If paper were a laser, the light that hits it would come out the other side brighter than it was before.

Since early in the twentieth century, such physicists as Niels Bohr, Louis V. De Broglie, Albert Einstein, and others learned how molecules and atoms absorb and emit light—or any other electromagnetic radiation—on the basis of the newly discovered laws of mathematical physics called quantum mechanics. One might say that when atoms or molecules absorb light, parts of them are wiggled back and forth or twirled with new, increased energy. Yet electrons in atoms and molecules store energy in very specific ways, with precise, discrete levels of energy. An atom or a molecule can exist either in a ground (lowest) energy state or any of a set of higher (quantum-defined) levels, but not in states between those levels. This means that they absorb light of certain wavelengths, and not others, because the wavelength of light determines the energy of its individual photons. (Some materials, because they have free electrons bound to no particular molecule or because they have so many different substances in them, do absorb and emit a continuous range of wavelengths, but that need not concern us here.) Figure 1 (top) illustrates the absorption of a photon (represented by the wavy line) by an atom (represented by the dot). As atoms or molecules drop from higher to lower energy levels, they emit photons of just the same wavelengths as those they are able to absorb; this is usually spontaneous, and this is the light normally emitted when molecules or atoms glow, as in a flourescent light bulb or neon lamp. Spontaneous emission of a photon is illustrated in figure 1 (middle).

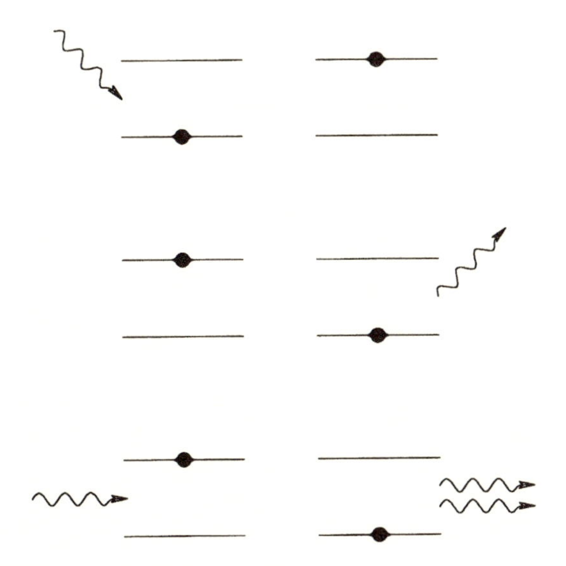

Figure 1. Stimulated emission of photons (bottom), the basis of laser operation, is contrasted schematically with the more usual absorption (top) and spontaneous emission (middle). When an atom in the "ground" state (black dot in top left) absorbs a photon (wavy arrow), it is excited, or raised to a higher energy state (dot in top right). The excited atom (middle left) may then radiate energy spontaneously, emitting a photon and reverting to the ground state (middle right). Also, an excited atom (bottom left) can be stimulated to emit a photon when it is struck by an approaching photon. Thus, in addition to the simulating photon, there is now a second photon of the same wavelength (bottom right), and the atom reverts to the ground state.

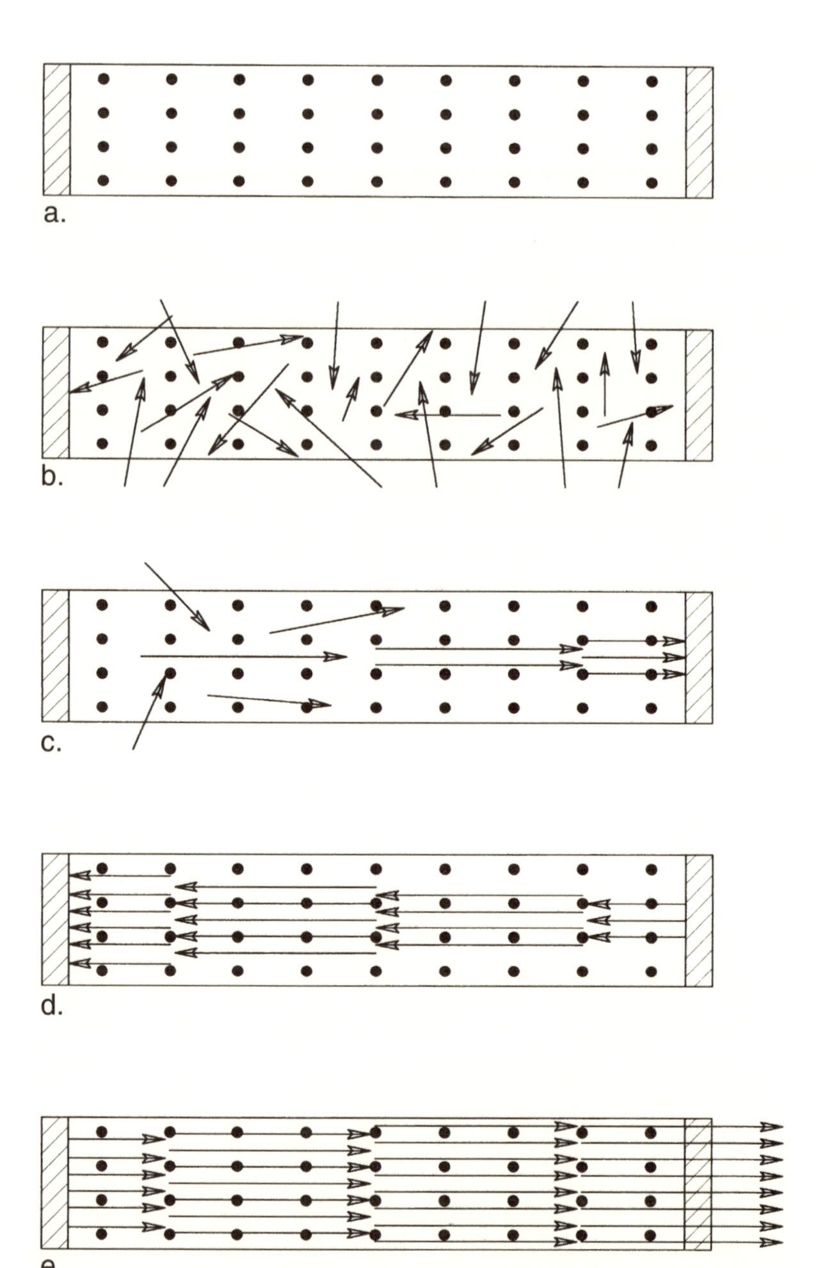

Figure 2. A laser using an optical quality crystal amplifies a light wave by stimulated emission, producing a cascade of photons. Before the cascade begins (*a*), the atoms in the laser crystal are in the ground state (black dots). Pumping light (black arrows in *b*) is absorbed and raises most of the atoms to the excited state (black dots). Although some photons pass out of the crystal, the cascade begins (*c*) when an excited atom spontaneously emits a photon (arrow parallel to the axis of the crystal). This photon stimulates another atom to contribute a second photon. This process continues (in *d* and *e*) as the photons are reflected back and forth between the ends of the crystal. The righthand end is only partially reflecting, and when the amplification is great enough, the beam passing out through the partially reflecting end of the crystal can be powerful.

Albert Einstein was the first to recognize clearly, from basic thermodynamics, that if photons can be absorbed by atoms and lift them to higher energy states, then it is necessary that light can also force an atom to give up its energy and drop down to a lower level. One photon hits the atom, and two come out. When this happens, the emitted photon takes off in precisely the same direction as the light that stimulated the energy loss, with the two waves exactly in step (or in the same "phase"). The result is called stimulated emission and results in coherent amplification; that is, amplification of a wave at exactly the same frequency and phase. This is illustrated by figure 1 (bottom).

Both absorption and stimulated emission can be going on at once. As the light comes along, it can thus excite some atoms that are in lower states into higher states and, at the same time, induce some of those in upper states to fall back down to lower states. If there are more atoms in the upper than in the lower state, more light is emitted than absorbed. In short, the light gets stronger. It comes out brighter than it went in.

The reason that light is usually absorbed in materials is simply that substances almost always have more atoms or molecules in lower states than in higher states: more photons are absorbed than are emitted. This is why one does not expect to shine a light through a piece of glass and see

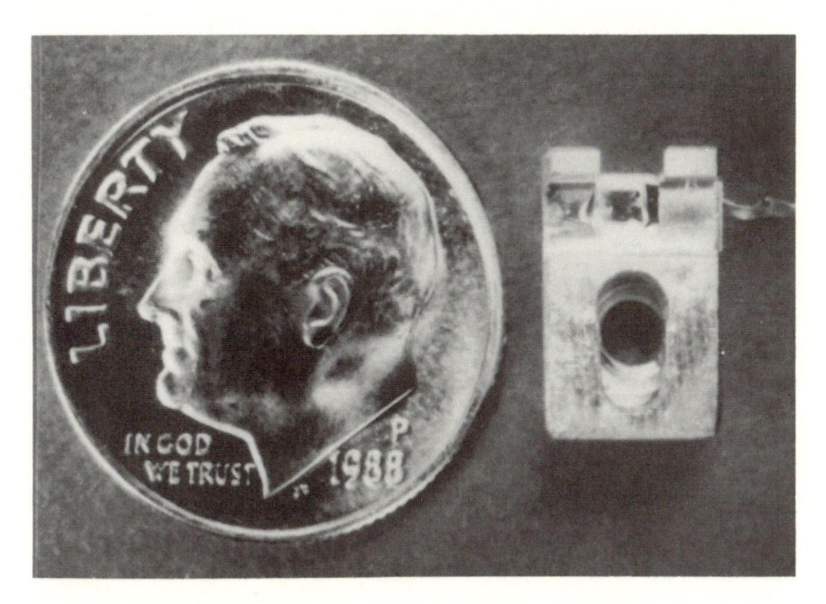

Figure 3. An early small laser (on right, as compared with U.S. dime), made of semiconducting material. It normally produces a small fraction, about 1/100, of a watt.

it come out the other side brighter than it went in. Yet this is precisely what happens with lasers.

The trick in making a laser is to produce a material in which the energies of the molecules or atoms have been put in a quite abnormal condition, with more molecules in excited states than in ground, or lower, states. A wave of electromagnetic energy of the proper frequency moving through such a peculiar substance will pick up rather than lose energy. The increase in photons represents amplification, or light amplification by stimulated emission of radiation. If the amplification is not very large on just one pass of the wave through the material, there are ways to beef it up. For instance, two mirrors—between which the light is reflected back and forth, with excited molecules (or atoms) between them—can build up the wave. Getting the highly directional laser beam out of the device is simply a matter of using two parallel mirrors, one of which is partially transparent, so that when the internally reflecting beam gets strong enough, a substantial amount of power shoots right on through one end of the device. These processes are illustrated by the diagrams of figure 2 that represent a laser in which atoms are excited to their upper-energy levels by a flash of light coming from outside the laser. This light penetrates the laser's transpar-

Figure 4. NOVA, which has been the world's most powerful and largest laser at the Lawrence Livermore National Laboratory in California. Its 10 laser beams can deliver 15-trillion watts of light in a pulse lasting 3-billionths of a second (45,000 joules). With a modified single beam, it has produced 1,250 trillion watts (1.25 petawatts) for a half of a trillionth of a second.

ent side walls, gives energy to the atoms, and prepares them for stimulated emission. A coherent beam of radiation emerges through a partially transparent mirror at one end of the device. At that point, one has a beam produced from light amplification by stimulated emission of radiation—a laser.

The way this manipulation of physical laws was discovered, and the many false starts and blind alleys on the way to its realization, is the story that follows. The story also describes my odyssey as a scientist and its unpredictable but perhaps natural path to the maser and laser. This is interwoven with the way the field grew, rapidly and strikingly, owing to a variety of important contributors, their cooperation and competitiveness—what might be considered scientific sociology. To be complete, the odyssey begins with my childhood.

2

PHYSICS, FURMAN, MOLECULES, AND ME

My childhood memories are chiefly of the 20-acre farm that my father, Henry Keith Townes, bought on the edge of Greenville, near the Blue Ridge Mountains in the northwest corner of South Carolina. Father's family had been farming in this region for generations, mainly growing cotton but also corn, other grains, apples, peaches, and sweet potatoes. My father's primary occupation was his law practice, but our family, like many in the area, followed a sort of Southern planter tradition, by owning a bit of productive land, farming it ourselves, or renting to tenant farmers. We kept a cow or two for our own use, and my father liked to see us kids working in the garden or out picking cotton, presumably to improve both our health and character. Greenville was a place of well-established sensibilities and rhythms. Like much of the South it was not at all prosperous, but it had a reassuring stability to it. Our own family of four brothers and sisters, then later two younger ones, was rather self-sufficient. We had a large garden, fruit trees and berries, cows we milked ourselves, chickens, ducks, and guinea fowl. We had a stream and fields to play in, plus a family tennis court that my father had built.

Life in such circumstances encourages one to pay attention to the natural world, to work with machinery, and to know how to solve practical problems and fix things innovatively, with what is on hand. Farms and small towns are good training grounds for experimental physics. I do not remember deciding to be a scientist, but I knew from a young age that I would be either that or, at least, a teacher of science.

Figure 5. With my older brother Henry (at left) and my sisters Mary and Ellen, I am exploring a Japanese umbrella, about 1920. My Aunt Clara, Mother's sister, had brought the umbrella and other curios back from Japan, where she was an early YWCA representative. We sit on the porch wall in front of the house in Greenville, South Carolina, where I lived until going off to graduate school.

I regard science in one way or another as the study of the universe, with the outdoors my first inspiration. My friends were primarily my older brother Henry, my cousins, and the lizards, birds, rocks, and insects around our home. Henry was a natural at biology, and I caught much of his enthusiasm. Our parents eventually got used to having snakes in cages outside the house and bits of foliage stashed in our bedrooms, as food for the caterpillars that we wanted to watch transform themselves into chrysalids and eventually butterflies, but which as often as not simply crawled off to explore the house. We had a large aquarium outside for fish, tadpoles, and turtles. Ours was a world of responsibility but also of adventure and imagination—some of my favorite books in our home dealt with self-reliance and living by one's wits and resourcefulness. Those included Johann Wyss's *The Swiss Family Robinson*, about a shipwrecked family that learns to build a new life on an island, and Ernest Thomson Seton's *Two Little Savages*, about a pair of youngsters who learn from an old man how to live off the land.

Most of our play had to do with practical things, building and exploring. I also liked to figure out how things worked. Recently, I came across a letter I wrote to my sister Mary when I was ten. It was mid-December, and I told her, "You asked me what I wanted for Christmas. I want mostly hardware so you better buy out a hardware store. I want some tin shears, some money to buy some iron and wood bits (as I want a particular size I had rather pick out my own), a flat file, a pair of glass cutters, some rifle shot and some one and two penny nails."

My brother and I were fairly competitive in what we did, but until I saw that letter I had forgotten one particular episode. In one passage I wrote to Mary, "Daddy has gotten a patten [*sic*] thing up that costs a nickel to patten anything. He did it because Henry fusses so much saying I copy him in everything." Apparently, I took after my older brother, and we competed so much that our father tried patent protection for the one who did something first.

On occasion, I got down to Charleston on the coast and visited the natural history collections in the museum there. I was fascinated by the differences in the plants and animals I saw on the shore plain and tidal inlets and those in my own Piedmont region. The geological contrast also impressed me. Back home, my brother and I did a great deal of walking and exploring, always watchful for birds, fishes, and other creatures. And I turned over stones, to see which creatures were underneath that were otherwise overlooked. Indoors, we had our hobbies, too. An uncle was dean of engineering at Clemson College, about 20 miles away. One day in the 1920s, he presented us with one of the early crystal radio sets, for which he had no more use, as he had bought a newer one. We tinkered around with it and listened in on the nation's first commercial radio station, KDKA, Pittsburgh. My father occasionally brought home, from a store he rented to a clock merchant, broken clocks for us to fiddle with, either to get them going again or simply to use their parts for whatever.

A memorable episode occurred one summer, during a visit to my grandmother's summer place in the mountains. From a branch of the Saluda River I netted a small, colorful fish. It looked like a type of minnow, but none of the standard guides to fishes showed this particular one. I pickled it in formaldehyde and sent it off to the Smithsonian Institution in Washington, D.C., with a letter asking the people there if they could identify it. I got a letter back with the information that this fish was either a new species or a previously unknown hybrid. It also asked me to please catch some more. Well, by that time I was not at Grandmother's place and I never had another chance to hunt or fish there. It is possible that nobody has caught that type of fish again. But it was thrilling that I might have come close to discovering a new species of fish and that Smithsonian scientists were interested!

My father was an amateur naturalist and would have made a fine scientist himself. However, when he was a youngster science was not a practical option, so he studied law. Our parents were strict about church, proper behavior, and school work. Any difficulty with school meant that they would drill us. Both of them went over our homework with us, and my father made Latin lessons fun as he rehearsed me with his own good memory of the language. The house was a simple but roomy, cedar-shingled, two-story affair that my father had built, after his first house on the lot had burned down, and it seemed full of encyclopedias. Father insisted on having reference books around; he also loved Mark Twain and Shakespeare. Mother, born Ellen Sumter Hard, was from Charleston and was rather intellectual as well—she took every course the local women's college offered, then took correspondence courses after that with a group of friends.

On my mother's side, we trace our ancestry back to Governor Bradford and the *Mayflower* colonists of Massachusetts. On my father's side, several siblings settled in the English colony of Virginia in the early 1700s, and his grandfather had moved into the Piedmont section of South Carolina later in the eighteenth century, about the time the Indians were being pushed out. The sister in this first Townes group married Christopher von Graffenried, a Swiss who in 1717 had founded an idealistic, communal city in North Carolina, called New Bern, after his native Bern. We were a blend of just about every Protestant denomination to come to America. We were Baptists but counted Lutherans, Methodists, Episcopalians, and Congregationalists in our sectarian background (there was no intermarriage back then between Protestants and Catholics!), and we were a blend of English, German, Scottish, Welsh, French Huguenot, and Scotch-Irish in ethnicity.

We did not have much money, but my family was proud of its roots and traditions. The South's defeat in the Civil War still reverberated in the early decades of this century. Among other aftereffects of the war in that part of the country was a cultural turning away from wealth as a source of social standing; there simply wasn't much wealth to be had in any case. Manufacturing and business were not prized occupations for many Southern families. People turned toward other values, such as family, personal character, tradition, church, the land, and learning. Perhaps because there was not much, the society seemed to save its pride by saying that money was not very important. Yet we had a sense that the South had once been a more important region, which had suffered a great loss. And history was a living thing to us. A treasured family tale was of my grandfather on my father's side, sitting on his uncle's knee and getting his eyewitness account of General Cornwallis surrendering his sword after the Revolutionary War battle of Yorktown. My mother's side had the story of a shipwreck near Charleston, which put her family ashore there, and how during the Civil War her family lost everything. They had guessed wrong about the path of Sherman's march to the sea and had stored their best possessions in Columbia, S.C., where they were destroyed or scattered as the Union army went through. We did not live in the past, but we knew its power and felt confident in ourselves because of it.

From all this, it should be no surprise that to grow up in our family was to regard education as a natural and automatic obligation. I was a fairly eager student, and when I became bored at school, my parents let me skip seventh grade. Our high school only went through eleventh grade. After the summer when I turned 16, I enrolled as a matter of course at the town college.

Furman University, a Baptist institution which then had only 500 students, may not seem to some an auspicious place to attend, but we felt it was a good place to be. I was in no hurry to be away from my parents. My father and older brother went to Furman; I went to Furman. Two of my sisters went to Winthrop, a nearby women's college, because Furman in those days was all male. My younger brother and sister broke away to attend Swarthmore, but we all felt that Furman deserved its place among the upper ranks of small colleges in the South, and with small classes it gave good personal attention. We also knew that, eventually, we could go to graduate school somewhere else, and could probably get scholarships or other help in paying the bills. While at Furman, I picked up a little money by tutoring, taking care of the Furman Museum, and selling apples from our farm.

Furman was not a place where the professors did much research in those days, but it had an intelligent faculty of high standards, and had given my father a good classical education. It looked the way a college should look, with ivy-covered walls and quiet walkways. Then, too, it was only about a mile and a half away, so I could save money by living at home.

Furman was small enough that I could know the professors fairly well and flexible enough for me to take the especially outstanding courses in any department. I could take almost all the really good courses that were given. This is one reason I wound up with a B.S. degree in physics and a second degree, a B.A. in modern languages, just as my brother Henry had done on top of a B.S. in biology. Another reason for the second degree was that I could have satisfied the requirements for a first degree in three years, but my parents felt I was too young to go off on my own, so I put in a fourth year. I didn't feel bad about staying at home for another year, although I was very ready to see new and different places. Overall, I have always felt Furman gave me an excellent and broad experience.

My career aims were not at first very specific. Hobbies in natural history with my older brother Henry made biology attractive. But he was so much better at it that, perhaps somewhat subconsciously, I eliminated it.

He went on, in fact, to be an entomologist. My younger brother George, who became a lawyer, frequently helped Henry collect the parasitic wasps known as ichneumon flies, which became his specialty. I also collected them for him throughout much of my career, whenever I went to some unusual part of the world. And natural history, along with the pleasures of hiking, scuba diving, and observing the beauties of nature have always been favorite hobbies for me.

After biology, mathematics was at first my top choice. It was taught by a somewhat distant relative and fine teacher, Marshall Earle. But things clarified in my sophomore year with my first physics course, taught by Hyden Toy Cox, Furman's physics professor. Physics, with its own wealth of mathematical logic, was appealing because it also seemed more connected to the real world than mathematics did. One could take just a few basic laws and build up a whole structure of theory and explanation that one just knew had to be right, the real way things work or close to it anyway. Biology was still largely descriptive, whereas physics could be precise, logical, and quantitative. That seemed to allow one to really understand just how things work. I took my studies seriously, seriously enough to get me into an embarrassing spot one day. I had worked hard on a problem in the textbook, finally seeing a way to get the answer to come out in agreement with what the book had said. The next day, Cox asked if I had gotten that problem, so I told him yes. "Now, how did you do that?" he asked. Of course, he knew that there was an error in the book and the given answer was wrong.

In those days before World War II, physics did not have much glamour. Most people had no idea at all what physics was. When my friends asked what it was, I had to explain it was sort of like chemistry, and sort of like electrical engineering, but somewhere in between. Professor Cox, who had a master's degree but no doctoral degree, was dedicated to his students. He also knew when to help and when to leave us alone, and I learned a great deal simply from textbooks.

During my junior year, the class did not get all the way through a physics text so I decided to study the rest of it myself during the summer. I vividly remember that summer, sitting on a moss-covered rock overlooking a stream near my grandmother's summer cottage in the Blue Ridge Mountains, with the section on special relativity opened on my lap, and reaching the startling conviction that Einstein had made a mistake in his logic. I went to lunch, and it was a heady few hours until I came back, sat down again with the book, and decided no, I was wrong and that Einstein had gotten it right after all. Despite that false alarm, it was an inspiring moment. It absolutely captivated me that, from a few simple equations, one could reach such profound and strange conclusions about the world. The idea that right there, on that page, was the explanation for why, at great speed, time must slow and an object must shrink in dimension but grow in mass, was a terribly exciting thing to confront.

In my senior year, my physics course consisted primarily of sitting down with a textbook, *Modern Physics* by G. E. M. Jauncey, and working every problem. A great deal of my early physics education came in this way, outside a classroom. A first encounter with electromagnetic theory came from an article by the famous physicist James Clerk Maxwell, in the *Encyclopaedia Britannica* in the Furman library. At the city public library were blue-bound technical journals that the Bell Telephone Laboratories published and gave free to public libraries. Those also were helpful. I recall well some fascinating articles in them that reviewed the field of nuclear physics, the hottest field of the time, written by a man named Karl Darrow of the Bell Telephone Labs. (I would have never guessed that some years later I would know him at Bell Labs, and be good friends with him when he was secretary of the American Physical Society and I was president.) I also ran across a fascinating short description of Karl Jansky's discovery in 1932 of radio waves from space, something which intrigued me immediately. It was my first brush with radio astronomy, a field that has played a large role in my career—and which has important intersections with maser and laser physics. What struck me about it was that here was a fascinating observation, and nobody had any idea about what generated those radio waves. Everything else I had encountered in physics by then had a theory or at least a hypothesis to go with it.

It was not all physics and languages by any means. Only four courses were required for the major in physics, and much of those consisted of working through textbooks. I also reactivated the campus museum, putting the biology collections back together, swam the quarter mile for the swimming team, and played trumpet in the football band. At one point I got it into my head to try for a Rhodes Scholarship. My father agreed, but warned me not to be too upset if I didn't get it. And, I didn't.

I was one of just two Furman physics graduates in 1935, which was roughly twice the school's usual output. I started looking around for a place where I could get a fellowship and eventually a Ph.D. One possibility was the University of North Carolina. A cousin, Earle Plyler, was a physics professor there and I told my father, "Well, since I have a cousin there, maybe I have a chance," but he said "No, no, no. Because he is your cousin, he'll lean over backward to avoid any semblance of special consideration." My father was probably right. At any rate, I didn't get any scholarship or fellowship offers there or at any of the several more famous universities to which I had applied. But I did get offered an assistantship at Duke University, so off I went there in the fall of 1935, and I soon chose to do research under the recently arrived Professor Woodbridge Constant. Duke was not then in the front rank of physics in the United States, but it had some good spectroscopy and cosmic-ray physics, enough to give me my first real taste of physics research. Constant had been at Cambridge University's Cavendish Laboratory in England for a year, and he was full of stories of some of the great names of the day, including Paul Dirac and Ernest Rutherford.

Constant had bought for Duke the components for two small Van de Graaff generators, early accelerators that used static electricity to give proton beams energy. The school expected him to use them to get a nuclear physics program started. Even a big Van de Graaff could only get its beams up to a few million electron volts, a tiny fraction of what modern machines provide, and Constant's were designed for only about half a million electron volts. But for the time, that was high energy.

It appeared that Constant did not have much taste for getting his hands dirty fooling around with machinery. At any rate, he had not been able to get the Van de Graaffs running well. So that is what I did, working out the physics and some theory that had not yet been recognized and writing it all up. This may be where my experience using and repairing things on the farm was a help. It also introduced me directly and indirectly to some other physicists with similar histories. To get a better idea about the ways Van de Graaffs might be used, I visited the Carnegie Institution in Washington, D.C. Merle Tuve, who ran that excellent Van de Graaff–nuclear physics program, had grown up in the farming country of South Dakota, right next door to another great experimental physicist, Ernest O. Lawrence. I perhaps had, and these other two men had, for sure, a feeling for how things work and an interest in figuring out, if something isn't working properly, how to fix it. The main use of a machine is to accomplish something. A person's cleverness in thinking of the right way of doing it is an important part of the game in practical life as well as in physics. Of course, there are some superb physicists who concentrate on theory only. There are many different tastes in this business. But I think a good combination of theory and experimental interest is, perhaps, the most powerful of all approaches to physics.

I had the Van de Graaff work done by the spring of 1936, which qualified me for the master's degree. Constant said, and this still amazes me, "Well, you know, yes, you have done your thesis. But really, we have never had anybody finish in one year and it just won't look too good. I think you'd better wait until later." Perhaps he was lacking in self-confidence. But I wasn't, so I asked if I could finish up and leave, and they could mail me the degree later. That's why my degree is dated 1937, a year later. All in all, Duke was pretty good for me. I took physics, mathematics, and chemistry, and I learned a bit of Russian, some of which I picked up from my roommate Cy Black, who had grown up in Bulgaria as the son of a missionary and learned Russian there. He was later to be a professor of Russian history at Princeton University, and best man at my wedding.

Caltech—From Low Point to High

One overt reason I left Duke so quickly was that I didn't get its one full-time fellowship in physics, for which I had applied. It went instead to a graduate student there who had come from Caltech, the California Institute of Technology. It made some sense—Caltech was famous for its physics, and Furman was not. For two years I had tried the Massachusetts Institute of Technology (MIT), and Cornell, Chicago, and Princeton universities and had been turned down for financial help at each. The reasons again were probably that Furman and Duke were not very well known in physics at the time. Finally, I decided to forget a fellowship or teaching assistantship and just find the best place that would have me and take it from there. I had saved up $500 and, with that, set my sights on Caltech.

In a sense, it was failure—the failure to get financial help at any among my first choices of graduate schools—that sent me off to Caltech. This was a failure for which I will always be grateful, a fortunate failure, because it made me go directly after what I really wanted.

Caltech, to my mind, was at the top of the physics world. After all, the man who beat me out for the Duke fellowship (but who, as it turned out, never finished his doctoral degree) was from Caltech. More important, Robert Millikan had made Caltech his, and Millikan was easily the country's best-known physicist, well known even to the public because of his Nobel Prize for measuring the charge on the electron. Not only that, but J. Robert Oppenheimer was oscillating between Caltech and Berkeley, with his students trooping back and forth with him, half the year at each place. Einstein spent some time there. On the faculty were Richard Chace Tolman, Fritz Zwicky, Carl Anderson who got a Nobel Prize in 1936 for discovery of the positron, and Linus Pauling, who was to have two Nobels. California seemed pretty exotic, too, with its palm trees and dry summers, the Pacific Ocean, high mountains, and vegetation so different from that of the eastern United States.

It was a long bus ride to Pasadena for me, sleeping in parks and in the bus. I stopped off in Alabama to visit an aunt, did some sightseeing in Texas, saw the Carlsbad Caverns in New Mexico, and, in a serious introduction to desert heat and thirst, hiked down to the bottom of the Grand Canyon and back up on two chocolate bars but, in my greenhorn ignorance, no water. It was summer, and I was sucking the pulp of cacti for moisture by the time I got back to the top of the canyon.

On the bulletin board at Caltech there were ads for places to stay. I found a house with a sleeping porch and space enough for two students, for $6 per month each. People with more money got the inside rooms, but the porch was fine with me. Everything I owned fitted in a small trunk. My porch mate was another physics student, Howland Bailey, who became a good friend.

By this time I had become fairly confident that I could do physics, but some catching up was clearly needed. One of the first required courses was electricity and magnetism, taught by W. R. Smythe. Both he and the course were tough. I remember spending 11 hours on three pages of the textbook trying to understand them adequately. I knew I did not have a lot of training. Yet when I worked at it, I found I could do it, and eventually I was solving problems that no one else in the class could do except for Leverett Davis, who became a good friend and a theoretical physicist who went on to be a professor at Caltech. Smythe took me on as a research student. He was working on a new textbook, and as his student I took on the job of working through all the problems, as he developed the book, and checking that his answers were right. I eventually did every problem, which meant that I learned a lot that year about electricity and magnetism. This was to stand me in good stead, as I later went into radar and then microwave physics.

Physics students in those days were a real mixed bag, because it was not as difficult as it is today to get into a good graduate school. The real problem was expenses (I did get a teaching assistantship at Caltech after a semester there, and so I was all fixed for the rest of my graduate work.)

Almost all the students were interesting to be with. They told good stories and had interesting and varied attitudes toward life, but some were not very good physicists. A few were extremely good, as were almost all members of the faculty. Willie Fowler had just finished his Ph.D. and was new on the faculty; he was obviously very bright. Youngish faculty members included Linus Pauling, whom I enjoyed and who was clearly very creative. Oppenheimer (known to us as "Oppie") was inspirational, and the students who migrated back and forth with him from Berkeley were impressive, including Leonard Schiff, Willis Lamb, Phil Morrison, and Bob Christy. George Volkoff, who with Oppenheimer worked out the theory of neutron stars, was a particularly good friend. We spent time bumming around together, taking hikes and trips.

Oppie did not take nonsense lightly, and could be cruelly cutting on occasion if he felt somebody was being stupid. His students adored him. It is probably true that he did not contribute as many major new ideas to physics as might have been expected from his remarkably sharp intellect, perhaps because it all simply came so easily to him that his attention was not fixed on one problem long enough. However, his classes were always about the latest thing. He knew quantum mechanics better, perhaps, than anyone else in the United States at that time, and he was amazingly quick in conversations.

Fritz Zwicky, the astrophysicist, was another interesting character, and very different from Oppenheimer. Oppie was forgetful about some things—there were a lot of stories of how he stood up young women for dates because he was working and just forgot all about them—yet Oppie was devoted to his students. Zwicky, who was Swiss and liked to go skiing and snowshoeing in the mountains, had different priorities. He would just head for the hills and tell us, "Well, read the book some and work some problems" and not worry about the class. Nevertheless, I learned a great deal from him about general thinking—approaches to problems through the use of very broad principles. Among these tactics was the use of dimensional analysis, an approach that leads one to focus on qualitative reasoning and on the units and quantities involved in a problem, rather than to wrestle with detailed theory. It sometimes is a quick route to a basic answer.

Caltech was not large in either number of students or campus size, and it was quite informal so there was interesting and healthy interdisciplinary interaction. For example, Linus Pauling, already head of the chemistry department, sat in on Richard Tolman's statistical mechanics course, which I was taking. He told Tolman "Well, it's been a while since I've taken any of your courses and I want to catch up on statistical mechanics." One of my physics student friends did his physics thesis on rockets, under the direction of the famous Theodore von Karman, who was head of aeronautical engineering at Caltech.

There was little government support for physics research in those days, with most of the financial grants coming from private sources, such as the Research Corporation, or from individuals; some of the money for nuclear physics at Caltech, for instance, came from the local Mudd family. There weren't very many physicists, either. That didn't seem to bother anybody—history shows, in fact, that it was an extremely rich period for the science, with refinement of quantum mechanics underway and with atomic and nuclear physics evolving rapidly. Physics was moving fast, invigorated by new ideas and many memorable characters. Nobody worried about policies to inspire more students to enter physics. We used to think it doesn't make too much difference how many students want to major in physics; only the good ones make a difference and will really do something; anybody who really loves physics and who wants to do it is going to be a physicist regard-

Figure 6. My laboratory and office at Caltech in 1938. The glassware against the right-hand wall are vacuum pumps involving boiling mercury; the elongated box in front of them is a prism spectrometer; and the tank contains compressed oxygen for blowing glass to make and repair the pumps.

less of public policy. I don't know whether such an attitude would be appropriate today, but it seemed to suit the times then.

I certainly had no doubts. At one point I came under a doctor's recommendation to give up physics, but it was unthinkable. My eyes kept bothering me. I consulted a specialist, who told me that all the reading I was doing was just too much, that I should plan on a different career. I compromised by setting my focus on experimental physics, figuring that working with instruments and equipment would be easier on my eyes than theoretical work. I don't know if that decision did my eyes any good but they still seem to be working fine today.

There were no women at Caltech in those days—a few years before I got there a woman had been admitted by accident, but that was it—so it was a somewhat monastic existence, yet it was all very pleasant. I joined the Pasadena Bach society—a choral group—and that had some women in it. I enjoyed the town. There was no smog then, and the local families were very friendly to Caltech students.

I also did some traveling, mostly in an old Dodge that my porch-mate, Howland Bailey, and I bought for $37.50. Several times I visited Berkeley, a beautiful place, and met with "Pan" Jenkins, a spectroscopist there (as dean a decade later, he offered me a faculty job, which I was not to accept until 1967, when I did join the University of California faculty).

With "How" Bailey I also visited Stanford, for a meeting of the American Physical Society. We slept with the Stanford family—that is, in our sleeping bags in the Stanford cemetery on campus. The campus cops woke us early in the morning to tell us we were not supposed to do that, but they did not really bother us. There was quite a bit of visiting back and forth between the Bay Area campuses and Caltech. Luis Alvarez, later a Nobel Laureate but then a postdoc in Berkeley, came down to see what I was doing because he was interested in isotope separation for the study of nuclei, and isotope separation was an important part of my own thesis.

We were of course all quite aware of world events. There was a feeling that war was coming in Europe and that it might draw us in. A few students were "marching for peace," as they put it, going around campus to demonstrate against what they regarded as warmongers and the commercial companies that would make money off a war. That this was happening at Caltech, a place very concentrated on its business of science and engineering, indicated how strong was the sense in America that war might really happen. But Caltech also had a firsthand view of what was happening in Europe. We all knew some of the refugees, particularly Jewish physicists and other scientists, who were fleeing the Nazis and coming to U.S. universities.

My combined lab and office was a room in the basement of the physics building, a large and solid stucco-finished structure, built after Millikan arrived from Chicago, and part of the main Caltech quadrangle. It was next to Smythe's office, close to the machine shop, and already filled, when I got there, with a rather elaborate apparatus involving about 30 glass vacuum pumps filled with mercury.

My thesis was on the separation of stable isotopes of oxygen, nitrogen, and carbon, and a determination of the nuclear spins of some of their rarer isotopes. The complex apparatus in my laboratory–office had been originally put together by Dean Wooldridge. As a student of Smythe's who had left just before I got there, Wooldridge had used it to separate some isotopes but had not gone on to measure their nuclear spins. Somewhat as I had done with the Van de Graaffs at Duke, I got the apparatus going again, made some modifications, and eventually managed to get some good physics out of it. For its time, the set-up was rather intricate. It had lots of glass tubing and about 30 gas flames boiling mercury in an equal number of pumps, which forced gas to diffuse through porous tubing. Lighter isotopes went through somewhat faster than heavier ones. This is the same strategy for separating isotopes that was used later at the Oak Ridge National Laboratory in Tennessee to enrich uranium isotopes.

To do its job, the system had to boil away night and day, for about three weeks, nonstop, without breaking, so I often got up during the night to attend it. Every once in a while one of the glass pumps would crack, or even break open, scattering mercury all over the floor, so I would clean up as much of the mercury as was practical, blow glass to patch up the pump, and start it going all over again.

Wooldridge had gone on to Bell Labs and would eventually be a cofounder of the Thompson-Ramo-Wooldridge (TRW) Corporation. I had no way of knowing it, working on that contraption, but he and I would later work together closely for several years.

I did get some useful isotopic samples out of the apparatus for spectroscopy. And I did my first spectroscopy on molecules, with a little background I had picked up from a short course on molecular spectroscopy in the chemistry department. With the samples I made of isotopically enriched oxygen, it was easy to show that oxygen-18 had zero spin, but this was already expected for theoretical reasons. Carbon-13 was the work's prime subject. I presented some results in a talk before a meeting of the American Physical Society and gave its nuclear spin as 1/2. This outcome, however, presented me with one of my first scientific dilemmas. After I left Caltech, Smythe went back through the data and prepared a paper, including my name as coauthor, that said the spin of carbon-13 is 3/2, in contrast with my earlier short abstract asserting a spin of 1/2. We disagreed over the strength of a particular line in the spectrum. He was judging the relative strengths of the lines by multiplying their heights by their widths, and the line that he felt was the critical one was broader than any of the others. I was judging intensity by the height only, suspecting that the unusually broad lines were actually superpositions of two or more weaker lines. But, I was by then on the East Coast and didn't know quite how to contradict my professor, so I let the paper come out as he wrote it. Later, the spin did turn out to be 1/2, so the published paper on my thesis is not exactly a world beater. However, I believe the episode taught me to be still more stubborn than I naturally was about sticking to my own conclusions. And the spectroscopy I learned was a small step toward the laser.

It was a time when I made a lot of friends, and really started to appreciate how the chance conversations and encounters of life lead, in totally unpredictable ways, to the events that shape a career.

Looking back now, I see that the chance of having a family such as I had, of having a brother with whom to go collecting in the woods and creeks, of having fine and encouraging professors and interactions with challenging young colleagues, all provide a lesson in how a career in science gets going. I mention this, because I am not at all sure that the public has a clear idea of how scientists get started and how they work. Eccentric scientists struggling alone with their ideas, brilliant social iconoclasts in isolation from everyday worries and following a clear internal vision, make popular drama but they are not the general rule. In fact a life in science, as with most things human, has haphazard aspects, taking off in directions that are hard to predict. The twists and turns depend as much on the friends and colleagues that one happens to make as on anything else. Now, this doesn't mean that one only needs connections to get ahead. A good scientist must have skills and diligence, must rely mainly and often stubbornly on his own judgment, and may spend long periods wrestling alone with problems, such as I did with those textbooks at Furman and Caltech. However, ideas, inspirations, and opportunities come as often from the people one happens to meet as they do from some sort of special vision. Any effort to chart a scientific career, or the evolution of a new concept or a new technology, must pay close attention to happenstance and to this collegial, interactive aspect of science.

3

BELL LABS AND RADAR, A (FORTUNATE) DETOUR FROM PHYSICS

In 1939, with my fresh Ph.D., it seemed absolutely clear that the ideal thing to do was to somehow work my way into a faculty position at a good university where I would be able to teach and do research.

Unfortunately, there just were not many jobs like that around. During the Great Depression of the 1930s, the research-oriented universities were hiring almost no new physicists. A number of new Caltech Ph.Ds were taking jobs in the oil industry to do exploratory work in the field. Such a position involved little research, but at least it provided a paycheck. The work included digging holes for oil field seismology studies. We joked that Ph.D. stood for "Post-hole Digger." Most of those who did get academic jobs had to settle for nonresearch-type institutions, including local colleges, which were glad to attract Caltech Ph.Ds.

AT&T's Bell Telephone Laboratories in New York had hired my predecessor Dean Wooldridge three years before—I was told he was in the first small group of physicists hired there since the 1929 stock market crash. My research advisor, Professor Smythe, put in a good word with some Bell people on their recruiting trip to Caltech, telling them that I was just as good as Wooldridge, who had already by then made an outstanding impression at Bell Labs. They asked me to fill out an application. I did it, but with little enthusiasm.

It wasn't long before Smythe got a call. The gist was: "What's the matter with this guy Townes? This is the most careless and messiest application we've ever seen. He just doesn't seem interested." Smythe had a quick talk with me. He told me I'd better get serious. Perhaps thanks in part to his plea on behalf of my soundness, I got an offer.

An industrial lab? I was not at all eager to help some company manufacture things or to make money. I hesitated. I felt I still had a good chance for a National Research Council Fellowship and with it at least a temporary position at Princeton. Smythe and Caltech astrophysicist Ira S. (Ike) Bowen sat me down. They said, "Look, this is a good job. Jobs are very scarce. You had better take it." But my goal was to be in an academic setting, an environment devoted to learning things primarily for the sake of learning them.

Among industrial research outfits, at least, Bell Labs had a better reputation in physics than just about any other such place. I knew of important physicists like C. J. Davisson, Lester Germer, Herbert Ives, and Harvey Fletcher, and their outstanding work at Bell Labs. And so, with some misgivings but the feeling of a sensible compromise, I took the job.

The money seemed quite good—$3,016 per year. I had friends who were taking jobs in other industries or who got teaching positions on the West Coast for around $1,800, and I was told that my salary would be the highest offered a new Caltech physicist, at least in recent years. I did not know it, but I was about to start a great adventure, encountering problems in physics and in intensive war work that honed my skills and shaped my career in singular, invaluable ways.

Not getting a first-class university job was a failure from which arose success, just as my failure to get a fellowship at Duke led me to richly rewarding years at Caltech. It is of course impossible to know ahead of time what failures are really successes in disguise, so the best thing to do is simply go after what seems to be the right thing at the time. It would be perverse to practice deliberate failure! Nevertheless, it is also valuable to know, when confronting a feeling of failure, that it could turn out remarkably well.

I was in no hurry and did not go straight to New York, where the Labs were at that time. Bell Labs provided $100 to get me and my belongings there by railroad. In those days, $100 could buy a lot of transportation, particularly if one rode buses. This was a chance not to be missed. Here was an opportunity to take a look at the geography, flora, and fauna of different regions. And, I wanted to pick up some Spanish. Mexico was nearby, and I had a Caltech friend in Mexico City. A Greyhound bus got me to Tucson, where for practically nothing I bought a third-class train ticket all the way to Mexico City.

I had with me an accordion I had bought from a German student, a rather ardent Nazi follower who spent a fair amount of time telling us all what a vital job Hitler was doing. I sat on that train in third class, on slatted-wood benches that were none too comfortable, and played a Nazi's accordion and sang songs with Mexican fruit pickers on their way home from the fields in the United States. It was a happy group. But I learned at meal time that third-class passengers were not admitted into the dining car. The Mexicans had plenty to eat, buying food that was handed in through the windows at each stop, but I worried about contamination and the famed "Montezuma's revenge." So, for the two days before we got into regions where fresh, peelable fruit was available, I lived on bottled beer. My relatives back in Greenville, many of whom would not even permit beer in the house, would perhaps have understood.

After Mexico City, I traveled on down to the Guatemala border and stopped there only because the bridge between Mexico and Guatemala was out. The return to Mexico City included an interesting couple of days in Acapulco, then a small and unspoiled seaside village, where I rented a little hut on the beach for 50 cents a night and swam in the warm water. People didn't use face masks in those days, there certainly was no SCUBA equipment, but I did some diving and could see enough of the sea life to ignite an interest in diving that was to stay with me. I got back to Texas on a Mexican train, continued on the bus to Greenville to see my family, and then went on to New York. The $100 from Bell Labs just about exactly covered the trip's total cost.

Bell Labs is in New Jersey now, but then it was on West and Bethune Streets in Greenwich Village, Manhattan. I rented a room nearby and made a plan to move every three months or so, just to learn the city. I like to keep moving and trying new things. New York is really a series of villages, and I wanted to get to know them. In addition to Greenwich Village, other places I lived were rooms near Columbia University, on Morningside Heights, and near the American Museum of Natural History. I would just put all my stuff in a trunk, get in a taxi, and move. By way of personal furnishings, I bought a collection of prints from the Metropolitan Museum of Art, framed some of them and put them up, changing them regularly. I took some voice lessons at the Juilliard School of Music too, then near Columbia, and sang in the choir at Riverside Church as well as in a church in Brooklyn.

The Bell Labs had been located for some time in what was, when I started work in September 1939, a rather old and undistinguished box-like building on the far west side of lower Manhattan. Another old industrial building across the street had also been taken over for lab use, and there was some further expansion into rental quarters in the rather more fancy Graybar building in the middle of lower Manhattan.

On my first day, I was taken in to see Harvey Fletcher, a distinguished figure in the field of acoustics and head of research in physics at Bell Labs. He was a kindly and fatherly man who explained that I was indeed to do basic physics research, with Dean Wooldridge as my immediate boss.

However, to get acquainted with the labs and select suitable work, I was to be given the unusual privilege of working for three months in each of four different research groups.

I had been hired to do basic physics and, for somewhat more than a year, that is what I did. Fletcher sent me first to work in magnetics, next in microwave generation, and then in electron emissions from surfaces. It was all very good physics. I believe it was the first time they had ever let anybody move around like that to get acquainted with the place before settling down. Things seemed very good. I could do research and, to some degree, follow my own instincts. But after actually only nine months, rather than the twelve originally planned, I was assigned to work under Dean Wooldridge on trying to understand secondary emission. This is the emission of electrons from surfaces bombarded with ions, and it had applications to gas discharge tubes.

The lab also started something that seemed fantastic at the time, a weekly get-together among eight or ten physicists and a couple of physical chemists. Bell Labs provided tea and cookies and told us just to discuss issues and concepts in the latest interesting research. This informal colloquium was, I believe, unique in industry then. The small group included Bill Shockley and Walter Brattain, who later were two of the three inventors of the transistor, Dean Wooldridge, Jim Fisk, later to be president of the Bell Telephone Laboratories, and Alan Holden, who was to collaborate with me after World War II in microwave spectroscopy. My new job plunged me into an environment more interesting and stimulating by far than anything I had expected.

Everyone who paid any attention to the news, and that included staffers at Bell Labs, knew that it was looking more and more likely that the United States would have to become involved in the war in Europe. The military in those days did not work nearly as closely with scientists as it did after World War II, but there was already war-related activity under way. One group of scientists and engineers was working on electronically guided anti-aircraft guns, an entirely new idea. Another group, I later learned, had somewhat earlier actually gone to the U.S. Navy Department and said, "We are experts in acoustics. We would be glad to help out. There must be problems with submarine detection and related underwater signals. Would you like us to try to help?" I was told the Navy replied with "No thank you, everything is under control. We know what we are doing, no further expertise is needed." That tune, needless to say, changed drastically as soon as German torpedoes began sinking our ships.

By the middle of 1940, the war work was picking up. One Friday, about a year and a half after I got there, I had a call from Mervin Kelly, director of research of the laboratory. Dean Wooldridge and I went to his office, meeting with him and Harvey Fletcher. He talked about the success the lab was having with electronically guided anti-aircraft guns, and said straight-away, "On Monday, I want you to start designing a radar bombing system by adapting the technology used for anti-aircraft guns and working with the Lab's radar people."

In my heart, I really objected to that. To be told, out of the blue, to drop physics and just start engineering was exactly what I had feared could happen at an industrial laboratory. But I knew circumstances were unusual. A war was coming on. Everybody had to pitch in. While I would have expected a more gentle exit from physics into war work, I knew it was not unreasonable for my participation to start then. I was also not very pleased to be doing military work. It was not that I thought the military was morally in the wrong position, but rather that it was a kind of dull and unattractive business. To be trying to think of ways to destroy things and kill people was not inspiring at all. However, it was by then early 1941, and I recognized the seriousness of the world situation. Wooldridge, the mathematically oriented engineer Sid Darlington, I, and a small group of technicians assigned to us, set to work the following Monday.

As our work proceeded, I got somewhat caught up in the challenges, and I wound up disappointed that none of the several radar bombing and navigation systems we came up with was actually used in the war. Somewhat simpler ones were. Ours were all fairly advanced, and they contributed to other designs that were eventually used in later aircraft, including early models of the B-52s built in the 1950s, descendants of which were used for some years by the U.S. Air Force. Yet the teamwork and testing provided a valuable education in how to run complex projects. Also, the work involved principles and inspired lines of thinking that were to be critical through much of my career, including the development of masers and lasers.

My personal life was also taking a new path. My second stop in exploring New York, after the apartment in Greenwich Village, was near Columbia University and the Juilliard School of Music. I lived in a small building where about half a dozen rooms were rented out to musicians—while I had a piano in my room, I was the only one who was not a full-time musician. There I met Frances Brown, a tall, slender young woman from New Hampshire, who had come to New York partly because, like me, she liked to explore new places. She had lived a year in Paris and a year in Florence and was quite sociable and outgoing. She had worked for a while at a fancy law firm, as a receptionist, and then gotten a job at the International House, near Columbia University, as director of activities. In the winter of 1940 she organized a ski trip, which had space for one more person. A dancer friend of mine, whom I had known in Pasadena, and who was going on the trip with her boyfriend, recruited me. So that is how we met. We had a good time on the trip. For some reason, there were a lot of Filipinos from the I-House learning to ski that week, so we helped out picking them up from the snow after they fell. We were married a year and a half later, in May of 1941.

Frances's family ran the Brown Company, a forestry and paper business in northern New Hampshire and Quebec. Before we were married, I had taken a trip with her immediate family into the forests of Quebec and she, like me, was very fond of the wild outdoors.

By the time Frances and I married, I was pretty deeply involved in the radar bombing project. We had built devices, and took them down to Florida for testing. First we flew out of Tampa, and then out of the Army Air Force base at Boca Raton and, for a time, out of the Naval Air station at Pensacola. Frances sometimes went with me and, except for one rather miserable and cold winter stay in a summer cottage near Tampa, we stayed at Delray Beach. By the time we were in Delray, our first child, Linda, had been born and could enjoy the beach. On those trips I learned a lot about working with the military.

One of the great Army Air Force secrets of the war was the Norden bombsight; the newspapers treated it as if it was magic. One read stories that with it our pilots could drop a bomb into a pickle barrel. Well, Bell Labs had an idea for a new radar-guided bombing system, and the Air Force asked that it be built. So we asked them, "What kind of accuracy do you need?" and they would only say, "Just get as much accuracy as you can." We then asked them what kind of accuracy they were already getting, so we could know what to shoot for. But that was a secret, you see. So we did just what they said and made a system as accurate as we could and as fast as was practical, which turned out to be about a year.

I recall in particular our first test flight, which was with an old bombing pilot, an Army Air Force colonel. We were to try dropping bombs full of sand on dummy targets, in this case on an old ship at anchor. On our very first bombing run, I was in the back of our B-24, tracking radar signals with the new equipment until the critical moment when the bomb was released. I dashed up to the nose of the plane, which was transparent plastic, to see how close we came. It missed by about 100 feet. The pilot exclaimed, "That's a damned good shot, if you ask me." This was our first clue as to what was really good. And it wasn't hitting a pickle barrel!

Late in the war we finally got a look at the Norden bombsight. The military wanted us to make a bombing system that would combine radar and visual sighting, so that either could be used. The Norden was nicely built, an opto-mechanical device, but even for that time it was rather primitive. It was never as good as it was popularly supposed to be. I think that was part of the secret.

The military generally kept its secrets fairly well, but most of us in technical work had a good idea about what the country was doing. I had friends who went to the Los Alamos National Laboratory in New Mexico. They dropped enough hints to let me know what the general state of progress was. In 1939, after the discovery of fission was announced, most of the physicists at Caltech realized that an atomic bomb might be possible. When everything in this area of research all of a sudden was hushed up, we knew it was because the government was starting a serious atomic bomb project. While publicly known measurements did not allow immediate assurance that a chain reaction could be managed, it was easy to calculate how powerful such a bomb might be. Enrico Fermi was among those who calculated how much uranium it would take to destroy, for instance, most of Manhattan Island. Despite the evident power of an atomic bomb, fear of possible world destruction by such weapons was not the issue. The overriding feeling was that the war had to be won by the Allies, but the great fear was that Germany might produce such a bomb first.

It was unusual then for men of my age not to be in the service. I vividly remember walking through Times Square with Walter McNair, then my and Dean Wooldridge's boss at Bell Labs. A complete stranger came up to me and said angrily, "You're not in uniform. Shame! A man your age ought to be in uniform, and helping out." McNair was worried that I might be upset, and he sought to reassure me that what I was doing was important and useful to the war effort. The fact was, despite my occasional frustration with the military and the management of the radar program, I knew perfectly well why I was not in uniform and that I was doing my part. Nonetheless, that episode illustrates how deeply the war then penetrated everyday American life.

So, I plunged into the radar work. It turned out to be a valuable time for me, in two ways. One was direct, and the other indirect. Workers at Bell Labs, and also those at MIT's Radiation Laboratory, followed up on the British invention of powerful magnetrons, whose pulsed signals made radar possible. Americans then worked toward still more power, shorter wavelengths, more precision, and greater sensitivity. Our first navigation-bombing system used magnetrons of 10-centimeter wavelength. But Bell Labs and the military were working intensively toward even shorter wavelengths. The goal was to improve directivity and sensitivity, while also permitting far smaller antennas—a big consideration when trying to fit a radar on an airplane. The military's failure to put any of the radar-guided bombing systems we developed into production was a continuing annoyance. Our first system worked well at 10 centimeters, and they said that it was a fine accomplishment, but how about one at 3 centimeters? When we did that, they told us to move to a 1.25-centimeter system. This constantly moving target was such a frustration that at one point I felt I should quit, to perhaps become a technical aide to General Joseph Stilwell, then operating out of China, or go somewhere else where I could contribute directly.

But I stayed, and the 1.25-centimeter radar project became an important episode in my development as a physicist. About ten years before that, there had been a unique study of absorption of microwaves by molecular gases. David Dennison at the University of Michigan predicted, from infrared studies of the ammonia molecule, that it would strongly absorb microwaves of wavelength near 1 centimeter. Claud Cleeton and Neil Williams at the University of Michigan set about to make a pioneering test, by using early magnetrons near this wavelength, ones which were much more primitive than the powerful type invented by the British. They simply got a big bag of ammonia, sent a wide spectrum of microwaves through, and found that they were most strongly absorbed in a band centered right at a wavelength of 1.25 centimeters. So here one could calculate what should happen, and then strikingly demonstrate it. The ammonia molecule is a beautifully simple system to understand dynamically. In it, a single nitrogen atom is sort of slung in a triangular trapeze formed by three hydrogen atoms. The nitrogen vibrates back and forth through the plane formed by the hydrogens. Geometrically, the molecule inverts itself repeatedly, much like an umbrella turning itself inside out and back again. The physics of the whole thing was fairly well understood, and infrared measurements had shown that the assembly should resonate or vibrate about 24 billion times a second, a frequency corresponding to microwaves with a 1.25-centimeter wavelength.

When the Air Force asked for a radar at this wavelength, there was not much chance of course that there would be enough ammonia in the air to interfere with it. Yet this early work showed clearly that molecular absorption could, in principle, be a problem. I really woke up when I saw a memorandum written by John Van Vleck of Harvard University that was being passed around among people working in radar. He pointed out that there should be an absorption by water molecules at about this wavelength, owing to a rotational resonance. He and Victor Weisskopf were also working at about that time on the theory of absorption-line widths. I was acquainted with their work, as well as with the experiments of Cleeton and Williams, and it seemed to me that any 1.25-centimeter radar needed further thought.

I started to focus on all this very carefully, and did a lot of talking with other physicists, visiting I. I. Rabi at Columbia, and consulting British radar experts. The affair produced mixed emotions in me. On the one hand, I found the theory behind microwave molecular absorption and spectroscopy increasingly exciting. On the other, I was more and more concerned that the radar was doomed. Part of me, in fact, perhaps wanted this to be the case, because I was still irritated that the military was not using the longer wavelength radars we had already developed.

On the theory end, one could show that as pressure in a gas goes down and collisions among molecules decrease, the widths of their absorption lines should get narrower and narrower. Well, here we were working with radar oscillators that could generate signals in this regime, and it became clear to me that some very precise molecular spectroscopy might be possible in rarefied gases. It was something I thought I might like to do as a scientist as soon as the war ended. The practical work of making radars might provide an opening into a whole new field of physics!

Meanwhile, I became more and more convinced that time and money was being wasted on a radar that would not work out in the Pacific, where the air is very humid and where our only remaining enemies would be, since the war in Europe was going well for us by that time. I told people at Bell, in the military, and in Washington of my concern about absorption by water vapor. I remember some character in Washington finally told me, "Well, you know, you may be right, but we've already made the decision, we can't stop now. You just might as well relax about it." They built the radar, and it didn't work. The beam would go a few miles and fade out, absorbed by water vapor. The radar system on which I had been working had to be scrapped—another failure that nonetheless provided me with a great boon. This intense work on the radar and microwaves is what oriented me toward molecular spectroscopy as a major career focus.

Already my work in Florida, flying radar-guided bombing systems, had contributed somewhat indirectly to my scientific career. This came in the form of spare time. When the weather was bad in Florida, or when test planes weren't flyable for one reason or another, I had time to think about physics. Ever since I was a student, I had been fascinated by the radio waves that Karl Jansky of Bell Labs had detected coming from all directions in space. Oddly, astronomers did not immediately pay much attention to those waves. Yet for me, it was an interesting problem to be solved, and I was dealing with radio waves. Something, some process that nobody yet really understood, was making radio waves at vast distances from Earth, filling the universe for all I knew. When my radar project was stalled, puzzling over this interesting and neglected problem seemed to me the ideal way of using my time for physics.

I did find a mechanism that seemed the likely source of such waves—electrons colliding with each other and with protons as a result of thermal energy in the ionized gases of interstellar space. After I completed a theoretical treatment of radiation resulting from such collisions and began talking about it with friends, one of them dug up a paper by the Dutch physicist Hendrik Kramers, which discussed production of X rays by a very similar mechanism as high energy electrons collided with nuclei. So, appropriate theory had already been worked out, but it had not been applied to this problem. I also located prior work by the American astronomers Louis Henyey and Phillip Keenan, who used this mechanism but with a somewhat erroneous theory—one which also overlooked Kramers's correct formulation from the field of X rays. Such confusion is not uncommon in a somewhat offbeat and neglected area. I was glad to try to straighten matters out and had fun publishing a discussion of what I believed was the correct formulation and its application to astronomy. My paper, as had the others trying to treat this subject, received little attention at the time. But it was essentially correct. I did miss another mechanism for generating some of those waves—the synchrotron process—in which charged particles emit radio waves, as magnetic fields force them into curved, spiral paths (a mechanism that the Russian scientists I. S. Shklovskii and V. L. Ginzburg later described).

Bell Labs was a particularly good place for planning an investigation of those waves from outer space. Not only had they been found by a Bell Labs man, but I was working with radar, a technology that used the world's most advanced radio receivers. The determination arose in my mind to get to work, as soon as war work eased up, on both molecular spectroscopy and the astrophysics of radio waves.

As the war wound down in 1945, Frances and I made a trip out west. I went to see Ike Bowen, one of my professors at Caltech who had urged me to take the job at Bell Labs and whom I trusted very much. He had been one of Millikan's associates, was very expert in optics and spectroscopy, and was then head of the Mount Wilson–Palomar Observatories—an outstanding astrophysicist. I went to him for sound advice about the most important things to try first with radio waves. We met in his office. He listened to me outline my hopes for expanding astronomy into radio frequencies, looked at me, and said, "Well, I'm very sorry to tell you, but I don't think radio waves are ever going to tell us anything about astronomy. I just do not think there is anything to do. The waves are too long . . . they are not directional, they can't really tell us anything."

Instinctively, I felt he must be wrong. I felt certain that radio waves offered a new tool which, when used and interpreted cleverly, could give some real information about their astronomical sources. But his remarks illustrate what often happens in any science. People in well-developed fields tend to be conservative, particularly with regard to ideas from outsiders.

As experts, they have a feeling they understand the field well and often do not much care for interlopers. In addition, their views of ideas or technologies behind new proposals outside their own fields of expertise are sometimes rather limited.

In fact, the United States missed the early development of radio astronomy rather badly. That took place first in England and Australia, largely due to work by radar people. It also developed early in the Netherlands due to the strong interest of astronomers there. At any rate, I put my radio astronomy and astrophysics ambitions on hold, perhaps in part because of what Bowen told me. I had to choose between microwave spectroscopy and astrophysics. I felt I knew more about the spectroscopy, so that is the path I took.

I asked the lab management to let me get back to physics. It was not quite that easy. Instead, I had to stay on about six months after the war ended, finishing up the radar work. My superiors also insisted that I get somebody to replace me before I could leave the project, which I did, recruiting my old porchmate from Caltech, Howland Bailey.

My determination to get back to physics did not make me popular. A friend told me that one of the vice-presidents felt I had become a good engineer, too good to go back to physics. Jim Fisk, then the head of the physics department at Bell Labs and later to become president, told me simply, "You've made a lot of people annoyed because you are talking about what you would like to do. You ought to be talking about what is good for the company." Bell Labs was a superb organization, but this kind of pressure was exactly why I had not wanted to go into industry. Naturally, I argued that what I wanted to do *was* good for the company. I maintained that the basic physics of molecules could be highly relevant to production of sophisticated communication equipment. This was not wrong, but the greater truth was I really did want to do physics simply because I wanted to do physics. I wanted to do what I felt was interesting and part of the fundamental process of discovery, and if I couldn't do it at Bell, then I would eventually have to go somewhere else.

In 1946 I wrote a report to Bell Labs giving all my reasons for pursuing microwave spectroscopy. The first paragraph read:

> Microwave radio has now been extended to such short wavelengths that it overlaps a region rich in molecular resonances, where quantum mechanical theory and spectroscopic techniques can provide aids to radio engineering. Resonant molecules may furnish a number of the circuit elements of future systems using electromagnetic waves shorter than 1 cm. Many difficulties in manufacture of conventional circuit elements for very short wavelengths can be obviated by using molecules, resonant elements provided by nature in

great variety and with the reproducibility inherent in molecular structure.

The report went on to outline ways that microwave spectroscopy and molecular resonances could be applied to signal detection, phase and attenuation control, and other uses. It went through my ideas that, at very low pressures, the line widths ought to be exceedingly narrow, including a proposal that the 1.25-centimeter ammonia line could be the basis for an extremely accurate frequency standard and what we now call an atomic clock.

Nothing in my report anticipated masers or lasers. In fact, I carefully noted that in view of the second law of thermodynamics, generation of intense short waves with molecules was not practical because reasonably intense radiation implied such high temperatures (to make the substance glow) that the molecules would separate into their individual atoms. My memo did, however, lay out a number of plausible ways that microwave spectroscopy could lead to useful technologies. It was enough to convince Lab management to let me work on it further, although my superiors were, in the long run, not convinced enough to allow the hiring of another physicist in the field, just a technician or two.

As I got busy with molecular spectroscopy, it turned out I wasn't the only one to turn wartime radar equipment to basic research. Other laboratories had magnetrons and klystrons—handy generators of microwaves—and plenty of waveguide equipment. Much of it was war surplus available at rock-bottom prices, and I bought a few klystrons from a sidewalk vendor in lower New York. The aborted 1.25-centimeter radar project was, in fact, valuable in contributing a great deal of surplus equipment to early postwar microwave research; I used some of it myself. So I should perhaps feel grateful that the establishment had been so stubborn about proceeding with 1.25-centimeter radar.

It did not take too long to get equipment working for spectroscopy. The predictions of extreme line-narrowing at very low pressure turned out to be correct, but I also found plenty of surprises. The ammonia-inversion line split into a number of lines in ways that didn't agree with available theory. In addition, each of those lines split into a hyperfine structure that reflected various orientations of the nitrogen nucleus. Those were exciting and challenging phenomena to investigate. I was forced to study all the subtle ways that various parts of the ammonia molecule might interact, and the way it might be distorted by outside influences, thus affecting the frequencies at which it absorbed microwaves. The many nuances of the molecule's behavior provided shapes to the measured spectral lines, and a forest of such lines, all much richer in detail than I had expected to find.

I was able to refine the theory of inversion, or turning inside out of the molecule, so that it would explain ammonia's behavior, and I also confirmed the general outline of the Van Vleck–Weisskopf theory of spectral-line broadening. My first paper on all this was in my out-box, ready for mailing and destined for publication when I received a manuscript from Oxford University. It was from Brebis Bleaney, who also worked on 1.25-centimeter radar during the war. He had not pumped the ammonia down nearly as far as we did, but he went far enough to see the line narrowing and get a somewhat coarse glimpse of the multiple lines. The two of us and, very slightly later but independently, William Good at the Westinghouse Electric Corporation opened the field of microwave spectroscopy. But as Bleaney had written up his paper first, I decided not to mail mine. I retrieved it, did further work on the structure and theory, and then sent off a more complete publication.

A number of years later, this particular action and timing became poignantly significant. Although Bleaney was to leave the field before long, I continued and developed many aspects of it into a prominent and rewarding new field of physics. One day much later, a member of the Nobel Committee visited me at Columbia University and asked about the history of microwave spectroscopy. I told him that Bleaney had published the first results, and I knew of them before I had published. A physicist friend told me that was unfortunate, for the Nobel Committee was considering a Nobel Prize for the field. Perhaps since I was not the clear-cut originator, and Bleaney did not pursue further extensive work, the field missed being recognized. I have always felt particularly bad about Bleaney, who has contributed importantly to physics. I myself was to have another chance.

I dove into study of accurate molecular structure and the spins and quadrupole moments of nuclei. The latter are related to the shape of nuclei. They produce very fine splittings of the molecular lines, because if a nucleus is not spherical, its different possible orientations in the molecule give the structure slightly different energies and frequencies. If one thought through the quadrupole spectral behavior, one could infer quite a bit about the shapes and hence structures of nuclei.

During the war, Bell Labs' radar work had interacted a fair amount with Columbia University, where work was being done on magnetrons. After the war ended, I continued to follow the work at Columbia; it was nearby and some of the physicists there shared my general interests in precise spectroscopy and the shapes of nuclei. I vividly recall going out to the Brookhaven National Laboratory on Long Island, New York, in 1947 for a conference on quadrupole moments. I. I. Rabi and Norman Ramsey, who had just moved on to Harvard, gave talks. After I spoke about my ideas on the effect of chemical bonds on variations in the energy of nonspherical nuclei (ones with a quadrupole moment), Rabi got up and said, "Charlie, that is a very pretty picture you are painting, but there is absolutely no science in it. There is just no scientific support for it."

I asked him for his specific objections, and he floundered. It seemed clear he was just not familiar with the wide range of information available about molecular properties. This episode illustrated the value of that weekly roundtable at Bell Labs. Wooldridge, Shockley, and I, along with others in the group, had together been through Linus Pauling's book on molecular bonds. I was pretty sure of the theory and felt I could answer all his objections.

Rabi did not relent that morning. He just closed the subject and wandered off. But in the afternoon we all went out to the beach for a swim. Rabi came up and asked me if I would like to come to work at Columbia. I had my first opportunity for a first-class university position! I did not take much time making up my mind. After my decision, Bell Labs generously let me continue there long enough to more or less gracefully finish the immediate research I had under way before I moved to Columbia.

4

COLUMBIA TO FRANKLIN PARK AND BEYOND

On January 1, 1948, I joined the Columbia University physics department as an associate professor with a $6,000 salary, only a bit lower than what I was being paid by Bell Labs. By then our second daughter, Ellen, had been born, but Frances and I thought we could get along in New York City on such a salary.

The department was busy putting itself back together after the disruption of the war. Some faculty members came back, and a few moved on. I filled a vacancy left by Norman Ramsey's departure to Harvard. Overall, I found a group with scientific interests perfectly in tune with my own.

It was not as large a physics department as some, such as Berkeley's, but it was first rate. I was pleased to be a member. Rabi was top man, both chairman of the department and head of the Columbia Radiation Laboratory, created during the war as a center for the generation of microwaves. I arrived shortly after the Radiation Laboratory had turned back to basic physics, including Rabi's research on molecular and atomic beams. Polykarp Kusch worked on magnetic moments of atoms and nuclei, using techniques similar to those of Rabi. Willis Lamb explored the energy levels of the hydrogen atom (work I had thought about trying myself, but Lamb had gotten to it first). A student there, Bill Nierenberg, had just finished an excellent thesis in spectroscopy of atomic nuclei, which was close to some of my own work.

That period in the Columbia physics department was to be even more remarkable than I could have realized at the time. When I arrived, I knew only that the department seemed interesting; it was both stimulating and comfortable for me. Yet only in retrospect do I fully appreciate the enormous productivity of the scientists then at Columbia.

During the 12 years I was a full-time member of the department, in addition to Rabi, Kusch, and Lamb, other professors there included T. D. Lee, Steve Weinberg, Leon Lederman, Jack Steinberger, Jim Rainwater, and Hideki Yukawa; all were to receive Nobel Prizes. Rabi was the only one so recognized when I arrived. Students during that period included Leon Cooper, Mel Schwartz, Val Fitch, Martin Perl, and Arno Penzias, my doctoral student who, in 1965, was codiscoverer (with Robert Wilson) of the cosmic background radiation (CBR), the relic photons from the big bang. All these were also to receive Nobel Prizes. Hans Bethe and Murray Gell-Mann were visiting professors there before receiving their Nobel Prizes. Then there were the young postdocs Aage Bohr, Carlo Rubbia, and my postdoc and close associate, Arthur Schawlow, now Nobel laureates.

Had I been a theorist, at my age—32—I should perhaps have been expected to have already done much of my best work. According to popular belief, theorists excel in their 20s. Experimentalists, however, have staying power. It takes time to get clever with instruments, and experience counts. The wartime detour had given me rich and crucial experience with electronics, with electromagnetic generators such as klystrons and magnetrons used in radar, and with practical engineering.

My research interests took no sudden new turn; the only change was the location of my lab and the opportunity to interact with students and a new group of interested senior physicists. The basic physics I did at Bell Labs, associated as it was with applied industrial and military work, gave me a good idea of what I wanted to do at Columbia. I was deeply fascinated with a number of lines of work which could be done with microwave spectroscopy. Those included highly precise examination of molecular structure, exploration of fundamental properties of nuclei, and improvements in the measurement of time. In addition, microwaves and radio spectroscopy provided new theoretical insights into the interactions of electromagnetic waves and matter.

Getting the equipment and team together was not a big problem. Today, a newly hired faculty member is often expected to bring along his or her own sources of support for a research program—and to find new ones as time goes by; usually this means a great deal of time putting together grant applications to government agencies. There are some advantages to today's competitive grants system. Nevertheless, writing proposals for supporting funds does drain time and energy from the lab.

The Columbia Radiation Laboratory had, by contrast, a secure source of support: a block grant from the joint services—the Army, Navy, and Air Force—administered by the Army Signal Corps. The grant's roots were based in the military's wartime interest in radar. Columbia's radar-related magnetron projects had led to wide-ranging physical research in the radio wave and microwave fields. Rabi told me about the workaday fashion in which the grants had been initiated and renewed. An Army representative came around one day shortly after the end of the war, took stock of all the magnetron work under way, and asked, "Would you be willing to keep this laboratory going, and do some physics with it? We can give you some money, something like a half million dollars a year." Rabi was very impressed by the generous offer, so much so that he told them half a million dollars was too much. They obligingly cut it down some!

The same kind of thing went on all around the country shortly after the war. The Office of Naval Research, under Alan Waterman from Yale, was the largest source of government money for basic science, and it was enormously important to research in U.S. universities. Waterman's philosophy, held also by such wartime research organizers as Vannevar Bush of MIT, was that a strong scientific establishment is important to the nation's future, important enough to the military specifically to justify such support. The result was a remarkably fruitful period of generous, open, and effective support of university research.

Military grants initially included considerable freedom over how to spend them. The system gradually became smaller and more rigid. This was partly because, somewhat later, the government set up the National Science Foundation (NSF), largely owing to Bush's inspiration; Waterman was the natural person to become its first director. Thereafter, the military could and did somewhat reduce its sponsorship of basic science. Still later, rules arose that required strict goals and accountability of military grants. Many universities began to find military-sponsored work to be more restrictive, and during the Vietnam war more politically burdensome, than it was worth. Yet in the first ten years or so after World War II, very few people had any reservations about doing civilian-type tasks with Pentagon money, and such support was a great boon to the nation's science.

In the same vein, in the late 1940s and in the 1950s, still fresh in people's minds were the important roles of the wartime Manhattan Project and radar programs. This gave physicists considerable social prestige, a great contrast to the years before the war, when most people never thought about physicists or physics at all. Suddenly, physics had an aura and physicists were popular at dinner parties. (The onus of nuclear dangers was to weigh on physicists heavily some time later, particularly in the 1960s and onward, after public interest took still another turn.)

We moved into an apartment at 120th Street and Morningside Heights with our two daughters, Linda and Ellen. Two more daughters, Carla and Holly, were born while I was at Columbia. Except for the worry Frances and I had over the extra care and attention it took to raise small children in the city, Columbia provided exactly the rich, challenging, and questioning environment I had sought. With weekly colloquia and other talks by faculty, visiting professors, and eager students, it was perfect. And before our fourth daughter Holly was born, we had managed to move into a house with a nice yard in the Spuyten Duyvil area of the Bronx.

My office on the tenth floor of Pupin, the physics building, looked out over the Columbia campus and, in the distance, the Empire State Building. When I got there my laboratory space—a large dusty room—was one floor up from my office. It was empty except for a big, man-sized metal box. The box provided a purely coincidental but very physical connection to my years at Bell Labs. It held a test apparatus that Columbia physicists had used to measure water vapor absorption of 1.25-centimeter radar waves. The tests were inspired by the doubts, associated with work on radar systems and already mentioned, that such waves could go far in moist air.

My immediate goals were to continue to extend microwave spectroscopy, including moves to shorter and shorter wavelengths. This would let me examine the behavior of many more molecules, plus the shapes, masses, and spins of nuclei. During my early work at Bell Labs, I had encountered some skepticism from other physicists that I was on a productive path. While ammonia had produced good results, several of my colleagues believed that its prominent microwave spectrum was such an oddity that it might be the only one we could usefully study. They simply doubted that there would be techniques sensitive enough to use in studying other molecules. Rabi also kept impressing on me his deep doubts about my theoretical interpretation of the effects of nuclear quadrupole moments, or shapes, on molecular spectra.

But microwave spectroscopy was already catching on before I arrived at Columbia. The chemistry departments at Berkeley and Harvard, plus the physics groups at MIT and at Duke and Oxford universities, were getting important results. Early work also was done at the industrial laboratories of General Electric, Westinghouse, and the Radio Corporation of America (RCA) in addition to my own work at Bell Labs. Within a few years we had found useful spectra in many gases. These included organic molecules, some diatomic molecules, quite a few salts that became gases at high temperatures, and a free radical, OH (water with a hydrogen atom torn off). An idea proposed by Bright Wilson, a chemist at Harvard, produced large improvements in sensitivity. With readily available klystrons and magnetrons—leftovers from the war—we found spectra that revealed new and interesting ways to determine the masses, the spins, and the shapes (quadrupole moments) of a variety of nuclei, as well as ways to get very detailed information about the structure, dipole moments, and interaction between molecules and electric fields.

Figure 7. My apparatus for measuring microwave spectra of molecules, built with my students at Columbia University, 1949.

By this time, a special abstract sort of friendship was arising from my research. Any devoted scientist develops a deep intimacy with the problems, concepts, or devices in his field. As for me, starting somewhat at Caltech, more and more at Bell Labs, and most richly at Columbia, my career brought growing familiarity and fascination with molecules. How molecules absorb and emit energy, their motions, and the behavior of their electrons and nuclei—all those things, while never actually seen by anyone, became real for me and easily visualized. When I try to figure out how a molecule behaves under particular circumstances, it seems almost like a friend whose habits I know. Ammonia, without a doubt, has been my favorite. Its simple arrangement of a single nitrogen and three hydrogen atoms has been pivotal in many important moments of my career. I have met this very familiar molecule in the insides of masers, as the mainspring of atomic clocks, in clouds among stars at great distances from Earth, and in the atmospheres surrounding some stars.

My molecular friend stood out in those highly productive first years at Columbia. This was partly because ammonia interacts strongly with microwaves. In addition, its unusual inversion (turning itself inside out like an umbrella in a high wind) and rotation, when combined with the effects of nitrogen and hydrogen nuclei, generated effects that demanded new theoretical explanations and careful measurement. Working with ammonia and other molecules in those early years at Columbia made it among the most satisfying periods of my career.

Not that I found myself in a perfect utopia. The military was still paying the freight, so there were gentle suggestions that I should do something in tune with the Pentagon's interest in magnetrons. Pushing spectroscopy to ever-shorter wavelengths or work with more difficult molecules did not quite fit the bill. Several times, Rabi or other members of the department asked me whether I really might like to work a bit on magnetrons. My reply was always no. The pressure never became intense, but it was there. For me, the magnetron was not particularly interesting in itself. A magnetron was merely a tool.

By no means did I entirely turn my back on the military or on any other agency that tried to change my research focus. While I do not believe I have ever permitted any such outsiders to make a real change in the direction of my research, I have also always sought to integrate any such contacts with my own genuine research interests. I have generally listened to outside agencies when approached and then done the things that seemed to me most worthwhile.

The military was not the only group outside the academic world interested in what we were doing. Private industry also paid attention. One unusual example came in the person of H. W. ("Hap") Schulz. A chemist at Union Carbide and Carbon Corporation, Schulz had been blinded as a result of a lab accident while a student at Columbia. He focused mostly on theory, including abstractly pondering the theoretical aspects of reactions.

One day in 1948, Schulz appeared in my lab with a proposition. I had never heard of him before, but I learned soon enough that he was a brilliant and inventive person. This visit was a lucky day for me. He had an idea for generating very specific molecular reactions with infrared radiation, radiation that has wavelengths shorter than 1 millimeter and occupies a portion of the electromagnetic spectrum between optical (visible) light and microwaves. Schulz noted that, if properly tuned, infrared could excite some molecules in particular ways, while directly causing no other changes in them. This selective alteration of molecular energies could provide a new way to control chemical reactions. In a mixture of reacting molecules, he hoped, infrared radiation could selectively speed up some chosen reactions. In principle, it was a very appealing idea. However, infrared wavelengths at suitable intensities were unavailable. Schulz had $10,000 from Union Carbide to offer to a knowledgeable scientist willing and able to try to produce such radiation. He asked me to give it a try. In effect, he had the $10,000 check right there in his hand. That was then a considerable amount of money—as much as scientists like me made in a year. It would be a great help in research. But I told Schulz that while my own interests certainly included development of waves still shorter than 1 millimeter, I saw no immediate way to do that. I told him it would be a bit false of me to take the money and to say I was going to do something for him with it. I did not want to interrupt my own work to make such an effort unless I felt I had a good idea to work on. At that, he left.

Within a couple of days, Schulz was back. He said he had been to other labs and decided he would simply like to give me the money anyway, because he thought what I was doing was impressive and was at least somewhat related to his initial aims for the money.

He had one question. What would I do with such a donation? To me, the answer was obvious. I would use $5,000 to finance a fellowship for a young postdoc to work with me, and the other $5,000 for research equipment. But of course I would continue to work on whatever I found interesting. Amazingly, this was fine with Schulz. I got the money, and it was renewed annually as long as I was at Columbia.

My first postdoc on the Carbide and Carbon fellowship was a young South African, Jan Loubser. He had trained at Oxford with my friend Brebis Bleaney, who was well known for microwave work. Jan worked on using the harmonics of magnetrons for molecular spectroscopy. The next year, the fellowship went to a highly recommended young optical spectroscopist from the University of Toronto. His name was Arthur Schawlow. He came to work with me to learn this new form of spectroscopy done with microwaves. Art was to become a major collaborator with me on many developments. Together we wrote a definitive text on microwave spectroscopy. He married my younger sister, and thus became a family member as well. And he was to play a key role in eventually making a revolutionary reality of Schulz's hopes for sources of intense infrared radiation—the laser.

Other visitors to the lab included, a few times every year, investigators from the military services. They routinely dropped in simply to take a look at what we were doing. One of these fellows was a Navy man, Commander Paul Johnson. He knew I wanted to study short wavelength microwaves and asked if I would consider forming an advisory committee to evaluate and stimulate work in the field of millimeter waves. This is the part of the spectrum just shorter in wavelength than the microwaves studied at the time, which were typically in the centimeter region or longer.

At that time, the Navy had no clear goal for millimeter waves. Johnson's committee proposal merely demonstrated that a few people in the Navy, familiar with the general directions of scientific research, wanted to be sure that no fruitful avenues for practical technologies were missed. The Navy and the military in general also knew for certain they did not want the United States to be surprised by enemy weapons that were based on techniques whose applications Americans had failed to recognize. It was an attitude toward science that contrasted sharply with feelings before the war, when most military organizations had been skeptical about needing some professor for advice on the way to go about the practical business of fighting wars. Yet during World War II, with its new radars, atomic bombs, ballistic missiles, jet aircraft, and electronic control and communication systems, the military developed a respect for science that it has never lost.

I had myself been stubbornly pursuing shorter and shorter wavelengths. Because they interacted more strongly with atoms and molecules, I was confident they would lead us to even more rewarding spectroscopy. Johnson's suggestion of a committee to consider ways that millimeter waves might be generated and used matched my professional interests quite well. He was not asking me to take up new projects in the lab, but only to put together a committee to look at potentials in a field in which I was interested already. I agreed, and the Navy gave Columbia money to finance the effort.

In early 1950, I sorted through names and asked senior people like John Slater and Al Hill at MIT and Leonard Schiff at Stanford University for advice on individuals at their institutions who might best contribute. The committee included only seven members, but they represented a wide spectrum of researchers. Besides myself there was John Pierce from Bell Labs, Marvin Chodorow from Stanford, Lou Smullin at MIT, and Andrew Haeff of the Naval Research Laboratory, all well-known microwave engineers. John Strong of Johns Hopkins University, a very important expert on infrared radiation, was on the team, as was John Daunt of Ohio State University, who worked on detection of infrared with low temperature devices.

We spent a good part of our time contacting industrial, government, and university laboratories to talk with people in the general field. At the committee's urging, Pierce wrote an article for the journal *Physics Today*, reviewing the importance of millimeter waves. We hoped to flush out, from physicists or engineers who read the article, concepts that we might miss on our own.

We racked our brains to think up ways to justify military support for research and development of millimeter waves. We didn't think of many. One drawback of millimeter waves, just like that of the 1.25-centimeter radar that took up so much of my time during the war, is that they don't go very far in the atmosphere. We tried to turn this shortcoming inside out, suggesting that it might actually be an advantage for shortrange communication. The dissipation of the signal would make things tough on eavesdroppers! We also noted that a wavelength of a few millimeters would allow antennas to be very small, a possible plus for packaging communication gear into aircraft or other vehicles. Space travel did not exist then, but we did note that in the low-density atmosphere of high altitudes, millimeter waves would not be absorbed and could therefore be very useful for communication, an advantage for very high-flying airplanes.

Primarily, we pondered how to produce the millimeter waves. One notion I had was that electron-spin resonances in very high magnetic fields would generate signals in the millimeter regime. That is, electrons would precess, or oscillate, as they interacted with the magnetic field at millimeter-wave frequencies. But I didn't see a very productive way to harness this process and we didn't pursue it very far. I had already in the laboratory worked on magnetron harmonics and on Cerenkov radiation to get millimeter waves. The latter involved electrons skimming the surface of magnetized materials.

The main problem, which Pierce pointed out in his article, was that generating millimeter waves by conventional means required a very small resonant cavity, only a wavelength or a small multiple of a wavelength in size. Making precise, delicate parts about a millimeter across is not easy, and to generate significant power one would have to pump considerable power through it, which wasn't easy. It would have to be strong and able to cope with a lot of heat. We did not in fact dig up any very great ideas for beating such problems. It was out of a sense of frustration over our lack of any substantial progress that the conceptual breakthrough finally came.

The Navy millimeter committee met several times in Washington, D.C. For one of our meetings, we chose April 26, 1951, to get together, because we could take advantage of an American Physical Society meeting at the same time. Art Schawlow was also there to present a paper. To save money, he and I shared a room at the Franklin Park Hotel. I felt that the committee was nearing the end of its useful life, and I was frustrated by our relatively meager list of new ideas or recommendations.

With small children at home, I was used to getting up early. Shortly after dawn on the day of the meeting, I awoke. So as not to disturb my late-sleeping bachelor companion—and future brother-in-law—I quietly dressed and crept out of the room to enjoy the early morning. In the back of my mind, I mused over the committee meeting and how we could be effective in our mission. In the fresh morning air of nearly deserted Franklin Park, I took a bench among beautiful red and white azaleas, in full bloom and laden with dew. It was a strikingly tranquil and lovely place, free of distractions. The problem turned over and over in my head. "Why is it that we just haven't been able to get anywhere? What is basically stopping us from being able to do this?" I asked myself. My mind first went back over the practical problems of tiny electron tubes or other resonators and of getting enough power through them to get decent signals out. As they had often done before, my thoughts drifted to a realm that was comfortable and natural for me, to physical mechanisms involving solids or molecules, especially the molecules to which I had devoted so much of my career and which I knew best.

The committee members and I knew from the start that some molecular transitions involve absorption and emissions at millimeter energies. I mused that, instead of trying to make small resonators, we really should somehow use molecular resonances, already provided by nature. I went through the usual arguments about why this would not work—that any collection of molecules would absorb more energy from a source of moderately high intensity than it would emit. One never shines a light or sends a signal through a gas, or anything else, and expects a stronger signal coming out! By heating a gas enough, one can make it radiate at many wavelengths, but to get the microwave portion of the spectrum radiating enough energy to be useful, the molecules would have to be so enormously hot that they would disintegrate into individual atoms. The requirement for an enormously high temperature results from a basic physical principle known as the second law of thermodynamics, which had always blocked me in previous forays down such avenues of thought.

I was well enough acquainted with the theory that radiation, even in a cool gas, can stimulate an excited atom or molecule to emit photons at exactly the same frequency as the stimulus, thus boosting the signal. This process is just the inverse of the absorption of radiation by an atom or molecule in a lower energy state; and for every atomic or molecular downward transition that contributes a photon to a passing wave, there are even more at lower energy levels to absorb the same energy photons in upward transitions. So, if a substance is in thermal equilibrium, this process is a net loser. That is what the second law of thermodynamics directly implies. The material soaks up more photons than it surrenders.

I cannot reassemble exactly the sequence of thought that pushed me past that conundrum, but the key revelation came in a rush: Now, wait a minute! The second law of thermodynamics assumes thermal equilibrium; but that doesn't really have to apply! There is a way to twist nature a bit.

Left to itself, as the second law describes, a collection of molecules does always have more members in lower energy states than in higher energy states, but there is no inviolable requirement that all systems be in thermal equilibrium. If one were, somehow, to have a collection entirely of excited molecules, then, in principle, there would be no limit to the amount of energy obtainable. The greater the density of excited atoms or molecules, and the longer the distance through them that the radiation wave goes, the more photons it would pick up and the stronger it would get. A few years earlier, I had even toyed with the idea of demonstrating this physical phenomenon but decided it was rather difficult to do and, because there was no reason to doubt its existence, I felt that nothing new would be proven by such a test.

On that morning in Franklin Park, the goal of boosting energy gave me an incentive to think more deeply about stimulated emission than I had before. How could one get such a nonequilibrium set up? Answers were actually well known; they had been in front of me and the physics community for decades. Rabi, right at Columbia, had been working with molecular and atomic beams (streams of gases) that he manipulated by deflecting atoms in excited states from those of lower energies. The result could be a beam enriched in excited atoms. At Harvard, Ed Purcell and Norman Ramsey had proposed a conceptual name to describe systems with such inverted populations; they had coined the term "negative temperature," to contrast with the positive temperatures, because these "negative" temperatures inverted the relative excess of lower-level over upper-level states in equilibrated systems.

It is perhaps a hackneyed device among dramatists to have a scientist scribble his thinking on the back of an envelope, but that is what I did. I took an envelope from my pocket to try to figure out how many molecules it would take to make an oscillator able to produce and amplify millimeter or submillimeter waves. All the required numbers about my friend, the ammonia molecule, were in my head. Ammonia appeared to be the most favorable medium. I quickly showed that we still needed a resonator, but now we would not have to pump electromagnetic energy into it. We could merely send a stream, or beam, of excited molecules through it, which would do the work! Any resonator has losses, so we would need a certain threshold number of molecules in the flow to keep the wave from dying out. Beyond that threshold, a wave would not only sustain itself bouncing back and forth, but it would gain energy with each pass. The power would be limited only by the rate at which molecules carried energy into the cavity.

A month earlier, I had coincidentally heard the German physicist Wolfgang Paul give a colloquium at Columbia. He described a new way to focus molecular beams, by using four electrically charged rods to form a quadrupole field—a field with four-fold symmetry. This system focused and intensified the beam better than Rabi's system with two flat plates, which had only a two-fold symmetry. Sitting on that bench, I calculated just how lossy the cavity resonator could be and still have oscillations produced by a beam of ammonia focused by Paul's method. That is, at what rate could the cavity lose energy because of imperfect conductivity and still allow an energy build up? The results indicated that it was just marginally possible the idea would work with ammonia. Using Paul's technique, one should be able to put enough excited ammonia molecules through a cavity to produce an oscillator that would operate at shorter wavelengths than could be achieved by any other known means. My calculations were for the production of electromagnetic waves in the far-infrared region, with wavelengths about ½ millimeter long that corresponded to the first rotational resonance of ammonia. But there was, in fact, no sharp limit that I saw to how short the wavelengths could be.

The signal, I also knew, would be coherent. The wave resonating in the cavity, reflecting back and forth, picking up strength from the molecules through which it passed, would maintain itself almost perfectly in phase and at a very nearly constant frequency, or wavelength.

I stuffed the envelope back in my pocket. In the hotel room, Art was up by the time I returned. I told him what I had just been thinking. He said, "That's interesting." He seemed only moderately impressed. To tell the truth, I also was not sure how far it would go. It would work in principle, but could we get one going, and if so how well?

At the millimeter-wave meeting, I did not bring up this particular idea. There was no fundamental reason that I could not have discussed it there and might possibly have developed it enough to mention in our report. I simply felt it was so new and fresh that I didn't want to take the chance that I had overlooked something; I needed to mull it over some more. I already knew that my concept contained no aspects or principles new to physics, but no one had considered such a scheme before, and it was not obviously easy.

When I got back to Columbia from Washington, I felt the idea for stimulated radiation was good enough that, first, I had to double-check the basics. I wanted to be completely certain that there were no mistakes in fundamental physics. I went and found my notes from a course in quantum mechanics that I had taken in 1939 at Caltech, from W. V. Houston. Yes, the equations there showed clearly that stimulated emission of radiation would permit amplification, and it indeed would produce a signal that was in phase, or coherent. The principles were all there in a course taught before the war to graduate students.

The characteristics of the resonant cavity also demanded more detailed thought. Waves needed to bounce back and forth in the cavity without losing energy too quickly to the cavity walls, a property described by what is called the quality factor, or Q, of the cavity. Calculation showed that, yes, a copper cavity could probably provide the needed Q (or low enough loss per bounce), even without lowering its temperature to achieve better conductivity characteristics. Most important, the calculation seemed to make it clear that I would not need to attempt the rather far-out possibil-

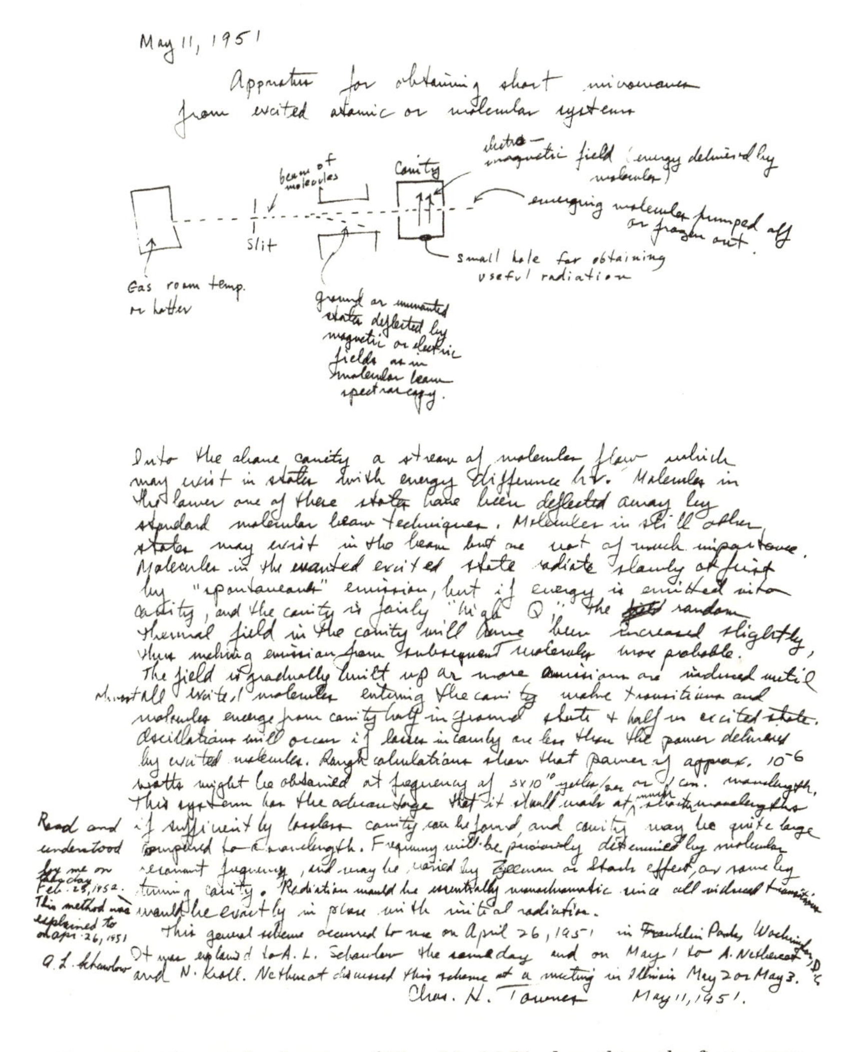

May 11, 1951

Apparatus for obtaining short microwaves from excited atomic or molecular systems

Into the above cavity a stream of molecules flow which may exist in states with energy difference hν. Molecules in the lower one of these states have been deflected away by standard molecular beam techniques. Molecules in still other states may exist in the beam but are not of much importance. Molecules in the wanted excited state radiate slowly at first by "spontaneous" emission, but if energy is emitted into cavity, and the cavity is fairly "high Q", the ~~field~~ random thermal field in the cavity will have been increased slightly, thus making emission from subsequent molecules more probable. The field is gradually built up as more emissions are induced until almost all excited molecules entering the cavity make transitions and molecules emerge from cavity half in ground state + half in excited state. Oscillations will occur if losses in cavity are less than the power delivered by excited molecules. Rough calculations show that power of approx. 10^{-6} watts might be obtained at frequency of 3×10^{10} cycles/sec or 1 cm. wavelength. This system has the advantage that it should work at much shorter wavelengths if sufficiently lossless cavity can be found, and cavity may be quite large compared to a wavelength. Frequency will be precisely determined by molecular resonant frequency, and may be varied by Zeeman or Stark effect, or some by tuning cavity. Radiation would be essentially monochromatic since all induced transitions would be exactly in phase with initial radiation.

This general scheme occurred to me on April 26, 1951 in Franklin Park, Washington, D.C. It was explained to A. L. Schawlow the same day and on May 1 to A. Nethercot and N. Kroll. Nethercot discussed this scheme at a meeting in Illinois May 2 or May 3.

Chas. H. Townes May 11, 1951.

Read and understood by me on this day Feb. 23, 1952. This method was explained to me Apr. 26, 1951
A. L. Schawlow

Figure 8. My notebook entry of May 11, 1951, describing the first maser idea—with a witness statement by Art Schawlow.

ity of using superconducting materials in the cavity; they would require very low temperatures and cause additional complications. Room temperature copper should do the job.

I satisfied myself that my idea employed only standard, known physics. In the years since, I have continued to learn the degree to which physics history was littered with hints that one could, in fact, use molecular and atomic transitions to make an amplifier. A few scientists suggested it explicitly; some had even undertaken experiments, though none seems to have thought of using a resonator and, in most cases, coherence of the amplification had been ignored. Here is a brief summary of these unappreciated clues.

Einstein in 1917 was first to explain how radiation could induce, or stimulate, still more radiation when it hits an atom or a molecule. The requirement is that the energy of the incoming photon, or quantum of radiation, approximately equals the energy that can be lost by an atom (or a molecule) that is making a transition from a higher to a lower energy state. Thus, Einstein described a process that would feed energy to the triggering radiation. He never considered coherence, but I feel sure that if asked, Einstein would have quickly concluded there must *be* coherence and that, if one had enough atoms in an appropriate upper state, one would get net amplification. Einstein, of course, was thinking of systems in thermal equilibrium. I don't have any evidence that he thought about nonequilibrium conditions. If prompted to think in such terms, he surely couldn't have missed the implications for amplification and coherence.

At Caltech, among my more interesting teachers was Richard Tolman, a physicist and chemist who worked in both general relativity and statistical mechanics. In 1924, he carefully discussed stimulated emission and absorption of radiation in a *Physical Review* article, calling the stimulated emission "negative absorption" (which is the same as amplification). For example, he wrote:

> The possibility arises, however, that molecules in the upper quantum state may return to the lower quantum state in such a way as to reinforce the primary beam by "negative absorption." . . . it will be pointed out that for absorption experiments as usually performed the amount of "negative absorption" can be neglected. (vol. 24, p. 697)

In one of his books of the period, he also commented that the stimulated emission must be coherent with the stimulating radiation, while at the same time he recognized that there was not yet any good mathematical proof of it. Thus the general ideas behind stimulated amplification were already fairly well understood in the mid-1920s by some physicists dealing with the quantum mechanics and the thermodynamics of radiation.

In 1932, Fritz Georg Houtermans, a German physicist, was told by an experimenter who studied gaseous discharges in his laboratory about the unusual intensity of a particular spectral line. Houtermans told me, long after the fact, that he remembered distinctly thinking that it might be a light (or photon) avalanche. This is a vivid way of envisioning stimulated emission in a gas that is out of thermal equilibrium. Houtermans said he didn't think about coherence and dropped the idea after another, more prosaic (and correct) explanation was found for the unusually bright emission line. More important, he evidently did not consider the phenomenon interesting enough to publish anything about it at the time, but he did publish his recollections of the event and the physics behind it after seeing the paper that Art Schawlow and I later wrote on "Optical Masers" and published in the Physical Review in 1958.

In 1939, a Russian, V. A. Fabrikant, published a rather obscure thesis describing the absorption and emission of light radiation in a gas in which he explicitly looked for what was called "negative absorption," or amplification. He did not discuss coherence or resonant cavities, but did suggest trying to obtain the necessary distribution of states by "collisions of the second kind," a mechanism now commonly used in gas lasers. However, he was not able to achieve any amplification, and his work was quickly forgotten. After our own maser idea was revealed, Fabrikant claimed a patentable version, dated June 18, 1951. That patent claim was published only in 1959, but I learned that Soviet law allowed patents to be rewritten and backdated. Fabrikant was definitely working on relevant concepts as early as 1939, but unfortunately he did not get very far and no one picked up his work as being particularly interesting.

A somewhat different type of energy-level inversion also got some study before that day in Franklin Park. A number of scientists had inverted nuclear-spin populations, using radio-wave excitation—though nowhere near enough to provide a net amplification, and had studied the results. This was the work that led the group at Harvard—Purcell, Ramsey, and Pound—to talk about "negative temperatures."

It was about 1948 when I had my own first idea to amplify radiation with a molecular beam, primarily as a demonstration of the physical principles. I did not couple it then with the idea of including a resonant cavity or of making an oscillator. A year or so later, John Wilson Trischka, a young postdoc at Columbia, got the same idea for demonstrating stimulated emission with a molecular beam containing inverted energy levels. He talked with me and worked at it on paper for some time before he decided that it would be very difficult and really would not prove much.

In 1950, Willis Lamb and Robert C. Retherford published a review of their study of the fine structure of the hydrogen spectrum. In describing the behavior of the atom they noted, in one sentence, that if the atom's higher energy levels were more populated than that at lower levels, "there will be net induced emission (negative absorption!)." They had not known that Tolman had already pointed out such a process, using the same terminology, as early as 1924. Willis was clearly aware of the process and understood phenomena such as coherence more completely than Tolman could in his day.

The Russian astrophysicist Vitaly Ginzburg has written of his professor, S. M. Levi:

> As far back as the 1930s Levi understood clearly the sense and the possible role of induced emission. Levi told me straightforwardly, "Create an overpopulation at higher atomic levels and you will obtain an amplifier; the whole trouble is that it is difficult to create a substantial overpopulation of levels." But why lasers were not created as far back as the 1920s, I do not understand. Much becomes obvious in hindsight.

Ideas about stimulated emission were thus floating around, but they were not being pursued. And although the basic physics was understood, no one had any idea that it could be useful. It seems clear that the study of microwave spectra of molecules was a key catalyst for putting together the various ideas needed for practical amplification or oscillation—probably because the needed concepts and skills emerged from a combination of practical experience with microwave engineering and devices and familiarity with quantum mechanical effects. At two other places in addition to our lab at Columbia, there were people working with such spectroscopy who had serious ideas about useful amplification by stimulated emission. One was Joe Weber, at the University of Maryland. He gave a talk to electrical engineers in 1952 about the possibility of amplifying by stimulated emission. He didn't have a cavity or a useful amount of amplification, but he certainly understood and suggested the basic idea.

In addition to Weber's short publication, and quite unknown to me until the spring of 1955, Alexander Prokhorov and Nicolai Basov at the Lebedev Institute in Moscow published a paper in 1954 describing how a molecular beam of alkali halide molecules might be passed through a cavity to produce a microwave oscillator. It was published somewhat after our publication on the ammonia maser, but it had been submitted before our paper came out. Their concept was remarkably like my own, except that they suggested using alkali halide molecules and a beam method that, as they showed, required a very high cavity quality, or Q, that was not very practical at the time. Because Weber and the Russians published after our system had been described in the Columbia Radiation Laboratory's quarterly progress reports, and since Weber visited me from time to time, one can wonder to what extent their ideas may have, perhaps subconsciously, been affected by our work. However, I see no reason to believe their work was not independent. The Russians proceeded toward building such a system and were later jointly awarded the 1964 Nobel Prize, with me, for contributions they made to the field.

By the 1950s, then, the idea of getting amplification by stimulated emission was already recognized here and there, but for one reason or another, nobody really saw the idea's potency or pushed it, except for me and the Russians, whose work was then unknown to me. A critical new idea that I added in Franklin Park was to use a resonant cavity so that the signal would go repeatedly through the gas, bouncing back and forth, picking up energy each time. This process would provide effectively infinite amplification, or oscillations limited in power only by the amount of molecular excitation that could be provided and the ability of the material to withstand so much energy passing back and forth through it. The whole plan only required properly putting together a number of ideas that were already known and floating around. Also critical was the recognition that this plan could be important.

The project to build a maser did not start immediately. I talked with my students and associates about it, but there was not even a lot of discussion. One of the few exceptions occurred when one of my postdocs, Arthur Nethercot, mentioned it at a meeting in Illinois in early May of 1951. He was sitting in the audience when he was asked to describe what was going on in short-wave studies at Columbia, so he stood up and reviewed the new concept for amplifying electromagnetic waves. This was the first public description of it, but except for Nethercot's memory, there was no record of it.

Essentially all of my research at Columbia was done with graduate students. To tackle the amplifier project I needed a good person, someone who was not only bright but who would have a few years to work at it. Jim Gordon had recently come from MIT, where he worked on atomic and molecular beams; and clearly he was a hardworking and quick student. "Here is an idea I think would be very interesting if we can make it work," I recall saying. "I can't be sure we can get it going on the time-scale needed for a thesis. Nevertheless, if you feel like working on it, I think it could pay off." Even if we could not get amplification out of it, I assured him, there were opportunities to do some very high-resolution spectroscopy with the apparatus, which would make for a fine doctoral thesis. The Carbide and Carbon Fellowship, which Hap Schulz had provided, allowed me to take on as the third Fellow another bright young postdoc, Herb Zeiger. He had done his thesis with Rabi in the field of molecular beams, and brought some of that expertise to the new project.

The general outline of the device seemed fairly clear. We needed a resonant cavity and an intense beam of excited molecules streaming through it. A key question was, what wavelength should we try first?

Because I was trying to get to shorter wavelengths, my first idea was to use ammonia with a rotational transition in the infrared region of the spectrum, for which the wavelengths—a few tenths of a millimeter—are much shorter than any that could be produced by other generators then available. There were many molecular transitions to choose from and such a generator, I felt, could be a real boon to further spectroscopy. It didn't take long, however, before we decided this was a hard nut to crack the first time out. We adopted the easier task of making a system work in the normal microwave region, for which we already had equipment and experience. We could then work our way down to shorter wavelengths and higher frequencies.

We set our aim on the strongest of ammonia's 1.25-centimeter transitions—one of the same group of ammonia resonances that Cleeton and Williams had measured back in the 1930s. This was the same wavelength region we had worked with for the radar that the wartime military establishment had wanted and which had been thwarted by water-vapor absorption. By the early 1950s, we knew quite a bit about the ammonia spectra, about techniques useful at this wavelength, and we had experience making good resonant cavities. Also, ammonia was my old friend. Today, it is almost laughable to think how easy it is to build a molecular or atomic amplifier, but the first one did not look so easy. Since this project was for a student thesis, it was also not done in a great hurry; it took more than two years to build up the equipment and work through the needed modifications and refinements.

The basic experiment involved a rectangular metal box, evacuated but with a tube leading in to introduce the ammonia gas. The gas diffused in through multiple small tubes to form a beam of molecules. Inside the box we mounted the molecular focuser, made of four parallel tubes about one inch in diameter and arrayed in a square. Downstream from this was the cavity, with a hole in each end to let the molecules go through. After the molecules flowed beyond the cavity, they were removed by either a vacuum pump or a surface at liquid nitrogen temperature, which condensed them as fast as they hit it. The box, with one side removed so that the four parallel tubes are visible, can be seen in the photograph of figure 9.

George Dousmanis, a student who had made his way to Columbia from the Greek Peloponnesus, also did some preliminary work with Jim Gordon. We all wrote the first formal plan for the device in the lab's quarterly report of December, 1951. This was not a public report in the usual sense; it was not, strictly speaking, a publication. Yet copies of it circulated around and we sent them to anybody who asked for them. Later, this quasi-publication would become an issue in battles over patents, but that was far from our minds at the time.

We were still working under a Joint Services contract, managed by the U.S. Army Signal Corps, and we had spent about $30,000 on the project during the first two years. I worked from time to time with Gordon and Zeiger, overseeing the project, as I did the other dozen or so graduate student theses I was supervising. We hardly rode a wave of encouragement.

When we showed the experiment to lab visitors, they would say "Oh, yes, interesting idea," and leave.

One day after we had been at it for about two years, Rabi and Kusch, the former and current chairmen of the department—both of them Nobel laureates for work with atomic and molecular beams, and both with a lot of weight behind their opinions—came into my office and sat down. They were worried. Their research depended on support from the same source as did mine. "Look," they said, "you should stop the work you are doing. It isn't going to work. You know it's not going to work. We know it's not going to work. You're wasting money. Just stop!"

The problem was that I was still an outsider to the field of molecular beams, as they saw it. It was their field and they did not think I fully appreciated the pertinent physics. Rabi was particularly assertive. On major scientific issues and debates, he was a wise and public-spirited man, but he could be combative over smaller arguments or day-to-day tasks. As far as he was concerned, he and Kusch were the beam experts, I was the molecular spectroscopist. Yet I had gone over the numbers very carefully with Gordon and Zeiger. I knew that the chances for quick success were somewhat marginal, but that the physics undergirding the concept was sound and the numbers promising. Rabi and Kusch, I felt, were going more on instinct. I simply told them that I thought it had a reasonable chance and that I would continue. I was then indeed thankful that I had come to Columbia with tenure.

After two years of work on the idea, Herb Zeiger's fellowship ran out and he took a job in solid-state physics at the Lincoln Laboratory. He told me later that in the meantime Professor Kusch had berated him for wasting two years on this hair-brained project, when he could have been publishing some solid papers in more conventional research areas. This apparently didn't divert Herb's thinking permanently, however, because at Lincoln Laboratory he was to later do pioneering work toward a semiconductor laser.

At Columbia, we kept working hard on our ammonia-beam device. Even though we had talked about the possibility of this new kind of oscillator and had had many visitors to the lab, and even though a few universities had put our internal laboratory reports on their open shelves, nobody else pursued the idea. It was not the only project in the lab, of course. I had a dozen Ph.D. students, quite a few for a physicist, to keep me busy with a variety of projects in microwave spectroscopy. But the oscillator got steady attention, because I felt it could provide a terrific tool for spectroscopic work.

The experiment was in my main laboratory room, so I saw Gordon and Zeiger frequently. We regularly talked over the design and plans. We worked particularly hard on improving the input of molecules and on making a cavity whose energy losses were less than the gains that would be picked up from the molecular beam. We started getting indications of stimulated emission almost as soon as we looked for it, and this provided some interesting spectroscopy of exceptionally high resolution. A good thesis for Jim Gordon was assured. The problem was that our cavities were a bit too lossy (too low in Q) to allow a resonant wave to build up as it bounced back and forth. We solved this problem little by little.

The idea all along was that, as we sent the molecular beam from one end of the cavity to the other, amplification would occur in microwaves reflecting back and forth across the cavity, more or less at right angles to the direction of flow in the molecular beam. Originally, we felt the whole cavity should be as enclosed as possible, but this was inconsistent with the job of getting molecules in and out. If holes in the cavity to let molecules through were too large, the resonating radiation might leak out the ends faster than it could build up. Adequately sealing the cavity to microwaves while getting molecules through it was a tricky problem. We kept trying different metal rings fitted into the necessary holes in the ends of the cavity. They could perhaps keep microwaves from escaping too fast while letting molecules pass. The solution, when it came, was simple. One day, Jim Gordon opened the ends almost completely; that was what put it over the top. Without a ring in each end, we could get plenty of molecules through the cavity. Our worry over too much radiation leaking from the ends was unnecessary. Apparently, without the rings, the pattern of radiation in the cavity became simpler and more efficiently confined. The cavity was quite long, and radiation largely bounced back and forth between the sidewalls, so not much leaked out the perfectly circular, large holes in the ends. Probably, the slitted rings, which had previously been fitted on the cavity, had neither been perfectly circular nor well enough connected to the cavity, thereby distorting the fundamental resonance pattern of radiation in the cavity and actually enhancing the loss of energy.

During a seminar with most of the rest of my students in early April of 1954, Jim Gordon burst in. He had skipped the seminar in order to complete a test with the open ends. It was working! We stopped the seminar and went to the lab to see the evidence for oscillation and to celebrate.

At lunch with my students some time later, as work on the new oscillator continued, I commented that we needed a name for the new device. We tried Latin and Greek names, but they seemed too long, so we settled on an acronym, based on the description: *m*icrowave *a*mplification by *s*timulated *e*mission of *r*adiation. The first "maser" had been born. This was about 3 months after Poly Kusch had insisted it would not work. When it

Figure 9. James Gordon (at right) and I were photographed with the second maser at Columbia University. The normally evacuated metal box where maser action occurred is opened up to show the four rods (quadrupole focuser) which sent excited molecules into a resonant cavity (the small cylinder to the right of the four rods). The microwaves that were generated emerged through the vertical copper waveguide near Jim Gordon. This second maser was essentially a duplicate of the first operating one, and it was built to examine the purity of maser signals, by allowing the two to beat together, thus producing a pure audio signal.

did work, he was gracious about it, commenting that he should have realized I probably knew more about what I was doing than he did.

This history—including the subsequent impact of the maser and its optical version, the laser—leads to an important point that must be in the forefront of any long-term scientific or technical planning. Some science historians, looking back on those days, have concluded that we were being in some way orchestrated, managed, manipulated, or maneuvered by the military, as though the Navy already expected explicit uses for millimeter waves and even anticipated something like the maser and laser. Politicians and planners, managing the budget, also generally believe plans must be focused by funding agencies on specifically useful directions. From our vantage point, the Navy didn't have any specific expectations at all about something like the maser or laser. Whatever new came out of the field was up to us. The military seemed quite uninterested in my maser work until some time after it was proven. What was critical was that I was free to work on what I thought was interesting and important. When one looks back in time, cause and effect sometimes get turned around. Industry and the military were important sources of generous support, but—in an experience shared by many academic scientists—throughout my career I have had to convince others, including sponsors, to let me keep following my own instincts and interests. Very often, this pays off.

5

MASER EXCITEMENT—AND A TIME FOR REFLECTION

Before—and even after—the maser worked, our description of its performance met with disbelief from highly respected physicists, even though no new physical principles were really involved. Their objections went much deeper than those that had led Rabi and Kusch to try to kill the project in its cradle; fully familiar with oscillators and molecular beams, these two never questioned the general idea. They just thought it was impractical and that it diverted departmental resources from basic physics and more sensible work.

Llewelyn H. Thomas, a noted Columbia theorist, told me that the maser flatly could not, due to basic physics principles, provide a pure frequency with the performance I predicted. So certain was he that he more or less refused to listen to my explanations. After it did work, he just stopped talking to me. A younger physicist in the department, even after the first successful operation of the device, bet me a bottle of scotch that it was not doing what we said it would (he paid up).

Shortly after we built a second maser and showed that the frequency was indeed remarkably pure, I visited Denmark and saw Niels Bohr, the great physicist and pioneer in the development of quantum mechanics. As we were walking along the street together, he quite naturally asked what I was doing. I described the maser and its performance. "But that is not possible," he exclaimed. I assured him it was. Similarly, at a cocktail party in Princeton, New Jersey, the Hungarian mathematician John von Neumann asked what I was working on. After I told him about the maser and the purity of its frequency, he declared, "That can't be right!" But it was, I replied, and told him it was already demonstrated.

Such protests were not offhand opinions concerning obscure aspects of physics; they came from the marrow of these men's bones. These were objections founded on principle—the uncertainty principle. The Heisenberg uncertainty principle is a central tenet of quantum mechanics, among the core achievements during the phenomenal burst of creativity in physics during the first half of the twentieth century. It is as vital a pillar in quantum theory as are Newton's laws in classical physics. As its name implies, it describes the impossibility of achieving absolute knowledge of all aspects of a system's condition. It means that there is a price to be paid if one attempts to measure or define one aspect of a specific particle or other object to very great exactness. One must pay by surrendering knowledge of, or control over, some other feature.

The most commonly encountered illustration of the uncertainty principle is the impossibility of learning both a particle's position and its momentum to unconstrained accuracy. The scientist must sacrifice one to get the other. The problem lies in the nature of the universe, not in the shortcomings of instruments. A corollary, on which the maser's doubters stumbled, is that one cannot measure an object's frequency (or energy) to great accuracy in an arbitrarily short time. Measurements made over a finite time automatically impose uncertainty on the frequency.

To many physicists steeped in the uncertainty principle, the maser's performance, at first blush, made no sense at all. Molecules spend so little time in the cavity of a maser, about one ten-thousandth of a second, that it seemed to those physicists impossible for the frequency of the radiation to also be narrowly confined. Yet that is exactly what we told them happened in the maser.

There is good reason, of course, that the uncertainty principle does not apply so simply here. The maser does not inform one about the energy or frequency of any specific, clearly identified molecule. When a molecule is stimulated to radiate (in contrast with being left to radiate spontaneously) it must produce exactly the same frequency as the stimulating radiation. In addition, the radiation in a maser oscillator represents the average of a large number of molecules working together. Each individual molecule remains anonymous, not accurately measured or tracked. The maser's precision arises from principles that mollify the apparent demands of the uncertainty principle.

Engineers, whose practical tasks up to that time almost never brought them face to face with such esoterica as the uncertainty principle, never had a hard time with the precise frequency the maser produced. They dealt all the time with oscillators and cavities, based on a wide variety of physical phenomena, which produced rather precise frequencies. They accepted as a matter of course that a maser oscillator might do what it did. What they were not so familiar with was the idea of stimulated emission, which gave the maser its amplifying power. Birth of the maser required a combination of instincts and knowledge from both engineering and physics. Physicists working in microwave and radio spectroscopy, which demanded engineering as well as physics skills, seem to have had the necessary knowledge and experience to both appreciate and understand the maser immediately. Rabi and Kusch, themselves in a similar field, for this reason accepted the basic physics readily. But for some others, it was startling.

I am not sure that I ever did convince Bohr. On that sidewalk in Denmark, he told me emphatically that if molecules zip through the maser so quickly, their emission lines must be broad. After I persisted, he said, "Oh, well, yes, maybe you are right," but my impression was that he was simply trying to be polite to a younger physicist. Von Neumann, after our first chat at that party in Princeton, wandered off and had another drink. In about 15 minutes, he was back. "Yes, you're right," he snapped. Clearly, he had seen the point. Von Neumann did seem very interested, and he asked me about the possibility of doing something like this at shorter wavelengths with semiconductors. Only later did I learn from his posthumous papers that he had already proposed—in a letter of September 19, 1953, to Edward Teller—producing a cascade of stimulated infrared radiation in semiconductors by exciting electrons, apparently with intense neutron-radiation bombardment. Along with his calculations, Von Neumann gave a summary of his idea:

> The essential fact still seems to be that one must maintain a thermodynamic disequilibrium for a time t_1 which is very long compared to the e-folding time t_2 of some autocatalytic process that can be voluntarily induced to accelerate the collapse of this disequilibrium. In our present case, the autocatalytic agent is light—in the near infrared, i.e., near 18000 Å [1.8 microns]. There may be much better physical embodiments than such a mechanism. I have not gone into questions of actual use, on which I do have ideas which would be practical, if the whole scheme made sense. . . .

His idea was almost a laser, but he had neither tried to use the coherent properties of stimulated emission nor thought of a reflecting cavity. There also seems to have been no reply from Teller, and the whole idea dropped from view. Later, in 1963, after the laser was well established, von Neumann's early thoughts and calculations were published; but by then von Neumann had died, and I never had an opportunity to explore with him his thoughts of 1953, about which he modestly kept quiet after we had the maser operating.

In the spring of 1954, the organizers of the Washington, D.C., meeting of the American Physical Society agreed to permit a postdeadline paper in which we described our new oscillator. Bill Nierenberg, who by then had left Columbia for Berkeley, told me later he recognized it as a very exciting development. Yet overall there was not a lot of immediate reaction. Our report was too late for the society's bulletin that described the proceedings of the meeting, so our first publication was early that summer in the Letters section of *The Physical Review*.

We started building a second maser almost immediately after the first one worked, in order to check the frequency of one against the other. We were joined by Tien Chuan Wang, a student from China with considerable engineering experience, and we had the second one operating in about 6 months. Each used the 1.25-centimeter transition in ammonia, with a frequency of about 24 billion cycles per second. Although they were essentially identical, they were not expected to have exactly the same frequencies. Slight differences in the dimensions of their resonant cavities could displace the two signals from each other by a tiny amount—by 1 part in 100 million or so. To test their constancy, we overlapped the outputs of the two masers so that they "beat" together. The signals came in and out of phase with each other at an audio frequency of a few hundred cycles per second. What resulted thus resembled, somewhat, the warble of a twin-propellered airplane, in which one engine is running just slightly faster than the other—the drone of one propeller alternately reinforcing, and then damping, the noise of the other. With our masers, the beat signal was very steady. Its pure sinusoidal form told us immediately that, indeed, both masers were operating at precise, nearly unvarying frequencies. If either of the maser's wavelength varied appreciably, the beat would have been noisy or irregular, but it was not. With data from this demonstration and other tests, we published in August 1955, a longer and more detailed paper on the maser in *The Physical Review*, which gave more complete information to other physicists on its intriguing properties.

As interest spread, we found ourselves with a steady stream of visitors. The Jet Propulsion Laboratory in Pasadena, especially keen on experimenting with this new device, sent Walter Higa to spend some time with us. We also got into a regular interaction with people at Varian Associates, Inc., near the Stanford campus in Palo Alto, who wanted to build commercial masers. By the late 1950s, after masers based on solids came along, so many papers on masers poured into *The Physical Review*, many of them speculative, that the editors declared a moratorium on maser publications! I believe this is the first and only time that journal has done such a thing. Maser research was so popular it became the butt of jokes. One making the rounds was that maser stood for "means of acquiring support for expensive research." It did help give us research support!

Even before Gordon, Zeiger, and I had gotten the first maser going, we realized that its steady, precise frequency would make it an ideal basis for an extremely accurate "atomic" clock. It was an obvious application, for the maser arrived on the scene with timekeeping already undergoing dramatic and rapid technological improvement. I had, in fact, worked with earlier types of so-called atomic clocks myself. Until that time, the best clocks, developed particularly at Bell Labs, used quartz crystals. Such crystals, however, gradually change their fundamental frequencies—in part because their mechanical vibration causes submicroscopic pieces of quartz to fly off. As a result, quartz clocks were good to only 1 part in about 100 million. This may seem impressive, but physicists wanted appreciably better accuracy than that. Several physicists had given the problem some thought. When I was still at Bell Labs, Rabi had suggested using the fixed wavelengths of radio-frequency transitions in molecular beams; and I had made an experimental "clock" based on a spectral line of ammonia.

Harold Lyons, at the U.S. National Bureau of Standards, was an electrical engineer with a good sense of basic physics and a particular enthusiasm for atomic clocks. He had enlisted my help as a consultant, and announced the first substantial and complete atomic clock in early 1949. Its accuracy was not much of an advance over the quartz-crystal clocks, but it was a move in the right direction and received a good public reaction. The radio broadcasting service of the International Communications Agency, called the *Voice of America*, as well as journalist Edward R. Murrow, and the U.S. secretary of commerce, to whom the Bureau of Standards reported, all played it up.

As the basis for a clock, the maser promised to provide the purest available frequency, at least over short periods of time—a promise it has fulfilled. As soon as the maser was working well, I let Lyons know that we had the perfect signal source. Such clocks are indeed accurate. The hydrogen maser, a later type invented by radio spectroscopist Norman Ramsey at Harvard, loses or gains only about one thirty-billionth of a second over an hour's time.

One must note that the maser is not the only good basis for such a clock. Another type of device provides a somewhat better average stability over a very long time period. That technique was developed during the 1950s by Jerrold Zacharias, an MIT physicist and former molecular beam colleague of Rabi's. His technique used a beam of cesium atoms without stimulated emission, and at present clocks of this general type provide the best long-term precision.

Historically, but somewhat inaccurately, the maser as well as cesium-atomic systems were all called atomic clocks, a term with great public appeal in the years shortly after World War II. With atomic bombs and atomic power in the news, an atomic clock seemed just the thing for keeping time. Of course the first maser, and the clocks stabilized on molecular lines, such as the one built by Lyons, are really molecular clocks. Nonetheless, the maser-based "atomic" clock, with its precision, was very satisfying to me for a deeply based reason. Very high precision physics has always appealed to me. The steady improvement in technologies that afford higher and higher precision has been a regular source of excitement and challenge during my career. In science, as in most things, whenever one looks at something more closely, new aspects almost always come into view. I could see that a clock built around a maser oscillator could be very useful; for example, in checking the precise rotational behavior of Earth or the motions of heavenly bodies. Precise timing would provide tests of relativity and its statements connecting rates of time and motion. Navigation and other practical fields would also profit from better timepieces. For the latter reasons, precise timing has been among the missions given to the National Bureau of Standards (later renamed the National Institute of Standards and Technology) and the U.S. Naval Observatory. Present-day global positioning systems (GPSs), which allow individuals with a small instrument to locate themselves within a few tens of feet anywhere on Earth or in the sky, are based on atomic timing.

My interaction and collaboration with Harold Lyons is just one illustration of the diverse, ever-surprising ways that relationships and friendships pay off in science. There is an unstructured, social aspect of science that is, I think, not sufficiently appreciated. By this I mean only that as developments and discoveries arise, scientists and their ideas are often thrown together, more or less by chance, or perhaps for reasons that at the time seem entirely utilitarian and single-purposed, which may pay important benefits in ways one could never anticipate.

The ripples of downstream consequence after Lyons' first involvement with me in 1948, shortly after I went to Columbia, were perhaps particularly chancy, but significant. In 1955, Lyons moved to the Hughes Research Laboratory in California to set up a group to work on spectroscopy and quantum electronics (a name we later coined for maser research and technology). He took with him from the Bureau of Standards some of the people with experience in microwave spectroscopy. And while at Hughes he hired an excellent group of physicists, including Ted Maiman, a man who, as we will see, was to have one of the starring roles in the development of the laser. Maiman, in turn, had recently finished a Ph.D. in radio and microwave spectroscopy with Willis Lamb, who was then at Stanford, after leaving Columbia and the radiation lab, where we had been close associates. A web of personal connections first spun at Columbia eventually spread across the nation. In science, there is usually no cold, objective inevitability to discovery or the accumulation of knowledge, no overarching logic that controls or determines events. There may be broad unavoidability to some discoveries, such as the maser, but not to their timing or exact sequences of progress. One has ideas, does experiments, meets people, seeks advice, calls old friends, runs into unexpected remarks, meets new people with new ideas, and in the process finds a career of shifts and often serendipitous meanders that may be rewarding and rich, but is seldom marked by guideposts glimpsed very far in advance. The development of the maser and laser, and their subsequent applications in my career and in science and technology generally, followed no script except to hew to the nature of humans groping to understand, to explore, and to create. As a striking example of how important technology applied to human interests can grow out of basic university research, the laser's development fits a general pattern. As is often the case, it was a pattern which could not possibly have been planned in advance.

What research planner, wanting a more intense light, would have started by studying molecules with microwaves? What industrialist, looking for new cutting and welding devices, or what doctor, wanting a new surgical tool as the laser has turned out to be, would have urged the study of microwave spectroscopy? The whole field of quantum electronics is almost a textbook example of broadly applicable technology growing unexpectedly out of basic research.

To return to the daily concerns of that time: our primary objective while working on the first maser was an oscillator with a high-frequency output. Not long after we had started work, I also realized that in addition to its use in spectroscopy, it would be a great amplifier. The maser can be several hundred times more sensitive than the old electronic amplifiers with which I had become so familiar while at Bell Labs. An amplifier, of course, is a device that has a small signal coming in one end, with a more powerful one coming out the other. The more sensitive it is, the weaker the starting signal may be and still come out cleanly in amplified form. Jim Gordon worked out, theoretically, many of the essential features of the maser oscillator, including its small fluctuations (approaches Art Schawlow and I later adapted to the laser). It would be a while, however, before we had a rigorous theoretical discussion of the maser's low-noise performance—that is, a precise statistical explanation of just how well it could amplify signals with little static or other clutter introduced during the amplification process.

In addition to using it for a variety of microwave spectroscopy studies in the year or two after the first maser was operational, I pondered how to extend the technology. The maser did a fine job demonstrating the principle, but as a useful tool it was severely limited. We needed masers that would work at shorter wavelengths, and also ones that could be tuned. The ammonia maser had an essentially fixed frequency, though several different ammonia-resonant frequencies might be chosen. An ideal generator for the spectroscopic study of atoms and molecules should provide signals tunable over a broad range of frequencies. One could then dial up and down through the generator's output range, probing for resonances in atoms and molecules and thus mapping their transitions and energy levels. For similar reasons the ammonia maser's value as an amplifier was limited. Many of my friends thought the ammonia maser was an interesting idea, but with such a narrow band, and no way to tune it, the thing seemed to them of little practical value other than for a clock.

I jotted down in notebooks at Columbia a number of ideas for tunable masers. Included was the notion of using a solid, rather than a flowing gas, as the masing medium. In many solids, when the electrons flip their direction of spin while embedded in an external magnetic field, they can emit or absorb microwaves. One then could imagine energizing the solid so most of its electrons were in one orientation. A wave of stimulated emission photons might then move through the solid, gaining energy as the electrons flipped to another orientation. The energy could be tuned by varying the external magnetic field, which would alter the energy released as the spins reversed direction.

During this period, I had a revealing encounter at a meeting of the Faraday Society in England in early April, 1955. To explain what happened then, however, I must first mention a much more recent incident, long after the maser and laser had been invented and had become entrenched throughout technological society. It occurred in 1991 at the 7th Interdisciplinary Laser Science Conference, in Monterey, California. I was there to introduce a session honoring two colleagues, Arthur Schawlow and Alexander ("Sasha") Prokhorov. Also attending the meeting was a Russian physicist from the General Physics Institute, established by the Soviet Academy of Sciences to honor Prokhorov, who was appointed its director. The representative of this institute gave me a beautiful brochure describing it, and I read the following:

> Specific mention should be made of quantum electronics which originated at the FIAN Oscillations Laboratory in 1954 when A. M. Prokhorov and N. G. Basov developed the first maser operating on a beam of ammonia molecules. The same discovery was made independently at the same time by C. H. Townes in the USA.

That they "developed the first maser" was of course blatantly wrong, even though Prokhorov, Basov, and I all would eventually share the Nobel

Prize for maser work, and they did, indeed, have good independent ideas in the field. The publicity that masers and lasers would receive has made it very tempting for public relations personnel of every institution involved to claim as much credit as possible for their own scientists. Some, including some institutions in the United States, were tempted too strongly.

I first met Prokhorov and Nikolai G. Basov at that Faraday Society meeting in England in 1955. I may have seen a few publications on microwave spectroscopy by Prokhorov before that, but I was not very aware of his work. What happened in Cambridge was an eye opener. It was, in those days, almost unheard of for Soviet scientists to visit the United States, but it was a little easier for them to go to Great Britain. So there they were. I had submitted a paper to the conference on our latest results with the ammonia maser, but had been told by the head of the conference that the subject did not fit their planned set of topics; so instead, I gave a talk on magnetic effects in molecules. The two Russians did not say ahead of time what they would talk about, which was a common Soviet way of doing things, because even their arrival was uncertain. When their turn came, they gave a discussion, to my amazement, on how an ammonia maser might work (though of course they did not call it a maser at that time). Their discussion was all theoretical, but they expected the device to work soon. After their presentation, I got up and said, "Well, that is very interesting, and we have one of these working." This also gave me a chance to describe our results briefly. It was only later that I learned Basov and Prokhorov had already published a paper about the possibility of such a device, and this is discussed below.

Being with them was great fun. We took a walk through the streets of Cambridge, where they seemed to talk a little more freely than at the site of the meeting. They were very eager to learn just how our maser worked, because they were still having trouble with their own. I was equally eager to hear about their work and situation. They had most of the same essentials, including a molecular beam and a resonant cavity. One thing that they had apparently missed was the quadrupole focuser, the scheme that I had picked up in 1951 from the German physicist Paul. Paul had published it in 1951, and so had we in 1954. Thus there was no compelling reason that the Russians would not have thought to use it, but they apparently had not. I told them everything I was doing, including the importance of the quadrupole focuser. They were scientists, good scientists, not secret agents, and I always openly shared and still share my university research with other scientists. Very soon, they had an ammonia maser working and even improved on the focuser, by putting in more than four poles. A number of later masers were to use their type of focuser, with many poles.

When Basov and Prokhorov wrote up the paper they presented at that 1955 Cambridge meeting, they properly acknowledged our already published work. They also referred to the paper that they had published in 1954, describing their ideas. I had not seen their paper before then, but of course I looked it up. It turned out that they did have a beam concept in some sense remarkably like ours, but initially using cesium fluoride and hence requiring resonators of a quality which was very difficult to achieve. It can never be known whether they had any inkling of our own internal Columbia University progress reports which, while they were not deliberately sent to the Soviet Union, were readily available to our sister laboratories and, for example, were on the shelf at Harvard's library. I personally have no doubt that they did come up with most of their ideas independently and as a natural development of their work in microwave spectroscopy.

Prokhorov and Basov are gentlemen and good scientists. I feel sure they themselves would not actually claim to have developed the first operating ammonia maser before we did at Columbia. Yet the affair illustrates the kind of pressures under which they operated in the Soviet Union, where so much stress had been placed on scientists to invent things first, to get ahead of the West and the United States in particular. And they were devoted and loyal Communists. The one time I remember differing strongly with them, considerably later, was on their view of Andrei Sakharov. They had signed a famous and very strong public statement against him, after he made public statements with anti-Communist impact.

It seems quite clear to me that the important concepts behind the maser had been coming together independently in Russia. Most interesting is that Prokhorov and Basov had strong backgrounds in microwave spectroscopy. Basov has said that they were thinking of molecule beams as a way of enhancing the signal-to-noise ratio in microwave spectroscopy and of narrowing the spectral-line widths. Those ideas then progressed toward stimulated emission, rather than absorption, and to making an oscillator. I do not believe it was a strange accident that maser ideas emerged from the fertile soil of microwave spectroscopy. In fact, as mentioned somewhat earlier, a third claim to originating the maser came from Joseph Weber of the University of Maryland. He, too, had a background in microwave spectroscopy and, at least for a while, felt he deserved credit for the first publication of the idea and hence its official initiation. (Details of that episode are discussed in the later chapter on patent issues.)

This stimulating encounter with the Russians came during an extremely fruitful time for me, yet it was about then that an important phase in my career was ending. While there was much excitement about the maser, I believed that the overall field of microwave spectroscopy was about done so far as physics was concerned. Most of the new physical ideas had been explored, both theoretically and experimentally. Microwave spectra had provided information about the spins and shapes of most of the atomic nuclei that could be easily studied in this way. The field, it seemed to me, from then on would be primarily of importance to chemistry. It was time for me to sign off and go on to something else.

Naturally, I took a lot of satisfaction in being among the originators of this field and staying with it to its maturity. However, I have never again approached any field in the same way that, as a young man, I immersed myself in microwave spectroscopy. Ever since, I have tried to devote my energies to opening new fields with the intention of moving on to other things as soon as they start to become well established. I like to turn over new stones to see what is under them. It is the most fun for me to be on the fringes, exploring aspects that seem interesting to me but have not attracted the attentions of others. Once a field is opened up and is successful, and once others are flocking to it, I feel my own efforts in it are no longer critical. At about that point, I like to move on to something I think may be promising but overlooked. Only with microwave spectroscopy did I hang around after a crowd had showed up.

To put an appropriate end to my engagement with microwave spectroscopy, I wrote a book on it with Art Schawlow, who by this time was my brother-in-law (he married my sister Aurelia in 1951, about 2 years after joining our group as a young postdoc). The book, published by McGraw-Hill, came out in the summer of 1955 and remains in print as a basic text in microwave spectroscopy.

After Art and I finished the book, and just after the meeting of the Faraday Society, I took a sabbatical. I had also just completed a 3-year term as Chairman of the Physics Department at Columbia and planned a sabbatical leave and trip abroad as a deliberate and significant break in my work. It was to be a time for reflection. My frame of mind was to be entirely open to new ideas or directions, to look around and see what was interesting. Radio astronomy, new types of spectroscopy, or high-precision measurements to test relativity and other basic physics were among the strong possibilities. As my intention was to break from the physics I had been doing up to that point, a crucial decision was whether to push hard on further development of the maser. The sabbatical was a fruitful one. The maser decision made itself. As it turned out, and somewhat to my surprise, the maser pursued me.

For support, I had a fairly unrestricted Guggenheim Fellowship, good for a little more than a year, plus a Fulbright Fellowship to cover teaching assignments at the University of Paris and at the University of Tokyo. To start, however, my idea was to just sort of cruise around Europe, visiting laboratories to see what physicists there were up to. It was also a good opportunity for our daughters, at least the older two, Linda and Ellen, to see some of the rest of the world. Our younger daughters, Carla and Holly, were 6 and 3 years old, respectively, and the visit to Europe didn't mean a lot to them. However, Japan was a great place for young children. The Japanese are very fond of children, and their many interesting customs, ceremonies, and toys made a lasting impression. During the summer of 1955, with our four daughters enrolled in camps in Switzerland and Sweden, Frances and I did cruise Europe, visiting interesting spots and scientific establishments. We took apartments here and there, including Innsbruck, Basel, Heidelberg, Cambridge, and Oxford (where I visited Brebis Bleaney, who had really been the first to do microwave spectroscopy on ammonia at low pressures and, thus, helped start the field). I gave a few seminars, talking about spectroscopy and, on occasion, the maser. I also sat in on as many scientific talks and seminars as I could.

I particularly looked into astronomy, a romance I had never quite kindled to that point, to consider how much I might contribute there. This included visits to the radio astronomy group in Cambridge, the big antenna at the University of Manchester, and discussions with the French radio astronomer, Evry Schatzman. Perhaps most significant of all, I was asked by Hendrik van der Hulst—the Dutch astronomer who had first suggested that the hydrogen microwave resonance at 21 centimeters wavelength might be seen in interstellar clouds—to give a talk at an international astronomy meeting on what other microwave resonances should be sought by astronomers. For the meeting, I discussed and wrote a paper on the molecules I believed might be found in interstellar space, a foreshadowing of my own later work in that field.

When the fall started, we all settled in Paris, where I had an office and taught at the École Normale Supérieure. In a Parisian bookstore, I saw the published Townes and Schawlow book on microwave spectroscopy for the first time. It symbolized for me completion of one adventure in physics and the open field ahead.

I spent the time in Paris at the superb laboratory of Alfred Kastler in the École Normale Supérieure, where he was doing a type of spectroscopy that was quite different from what I had done myself. Many of his students of that time are now well known, and keeping up with their careers has been a pleasure since then. By coincidence, I shared an office with a bright young graduate student named Claude Cohen-Tannoudji. Some time later, not only did Alfred Kastler win a Nobel Prize for work he had already then done in spectroscopy, but so also did Claude. He became an important figure at the École Normale Supérieure and won the prize for cooling atoms to extraordinarily low temperatures with a laser. Another coincidence, of more obvious significance at the time, was that a former microwave spectroscopy student of mine, Arnold Honig, was there too, working with physicist Jean Combrisson. Arnie told me something very interesting. He and Combrisson had just discovered a silicon-based semiconductor in which the electron-spin resonances—the energy levels of electrons in a magnetic field—had a very long relaxation time. This means that, in this material, energetic electrons tended to last a long time in a high-energy state, with their spin oriented in one direction with respect to an applied magnetic field, which gave them a high energy before they flipped over to a lower energy state, with their spins pointed the other way. Their relaxation times were as long as 30 seconds, and this seemed an eternity compared to most other materials I knew about, whose relaxation times are typically only a few hundredths to thousandths of a second.

I had already been thinking about how to make a maser with electron-spin resonance in a solid, but had laid the idea aside because it seemed rather difficult. With this news, I immediately said to Arnie, "Hey, that's just what is needed for a good maser amplifier." Here was a substance that not only would stay in the upper state for a long time, primed for the stimulation of radiation, but its frequency could be tuned simply by varying a magnetic field. It could make the needed tunable amplifier. Honig, Combrisson, and I agreed that we should work on it together.

That winter, I briefly went back to the United States to attend some scientific meetings. Encouraged by telephone conversations from Paris to my friends at Bell Labs, I took a side trip there. My former student Jim Gordon had by then joined Bell Labs, and I believe he was the first person I had called. He was, of course, very interested in anything to do with masers. I told him and George Feher, a Bell physicist who had done a lot of work on electron spins, about our thoughts in Paris, and that we needed some specially made silicon material. Today, many types of semiconductors can be gotten from a variety of electronics companies, but back then Bell Labs, with its excellent solid-state physics and transistor programs, was almost unique in its ability to make especially good semiconductors, and such materials were scarce.

Bell Labs provided just the material we needed. I took it back with me to Paris, where Honig, Combrisson, and I could pursue work on a tunable, solid-state maser. We had just three months to work on it before I had to leave for Tokyo. Gordon and Feher had agreed to wait a few months before working on the solid-state maser themselves so as not to take our idea and compete with us. We never got net amplification in Paris—not enough to overcome inevitable circuit losses—but we did get substantial enough amplification to show that it could work with a little more development. Before I left Paris, we wrote up our idea and quick results. To recognize

France properly, we published it in French in the classic French journal *Comptes Rendues*.

Back at Bell Labs, while my family and I moved to Tokyo for further adventures, Gordon and Feher honored their agreement to wait a bit before trying out the same material and then pushed to make the first solid-state maser. I saw their results some time later, which indicated that they did, just marginally, get over the mark of net amplification.

We were not the only ones working along these lines. Sometime early in 1956, a bit after we began work in Paris, but apparently independently, the idea to use electron spins for maser amplification occurred to Woody Strandberg, an outstanding microwave spectroscopist at MIT. He presented an enthusiastic talk about it, with Nico Bloembergen of Harvard in the audience. At the end of the talk Nico asked Strandberg why anybody would want such a device. Strandberg enlightened him to the potential of masers as amplifiers with sensitivities far beyond those of any other existing amplifiers. Soon after that, Bloembergen saw our new Paris paper.

Bloembergen was a radio and microwave spectroscopist of liquids and solids. He had worked for some time with paramagnetic materials (that is, substances with a modest response to external magnetic fields), and he knew a great deal about the electron spin behavior in such materials. It was perhaps this considerable experience that led him to envisage a clever system, one still better than what we had worked on in Paris. Thus was born the three-level electron-spin maser. In it, microwave energy first boosts electrons from their lowest level to the third level up, bypassing a second or intermediate level. Thus energized, the material can amplify a somewhat lower frequency of input radiation, one corresponding to the energy released as the electrons fall to the intermediate level. This meant that one could pump the maser medium to its excited state with energy of a different wavelength than the maser's output.

For such a maser to operate, its electrons have to be in a crystal with a particular type of asymmetry. As I was not very familiar with paramagnetic materials at the time, I had not even realized that this arrangement of energy levels was possible in such materials. Bloembergen was better versed than I was in this field, and he envisioned the three-level paramagnetic maser.

The ferment of ideas, with its interplay back and forth among scientists, was soon to become still a bit more complex. H. E. D. Scovil, a Bell Labs physicist who had worked on paramagnetic resonances with Brebis Bleaney at Oxford, had also been led, by my friends Jim Gordon and George Feher, to an interest in electron-spin masers. He learned that Bloembergen had a new idea for a maser, but Bloembergen was keeping it secret for the time being. Spurred perhaps by the bare knowledge that something new was in the air, Scovil also came up with the three-level electron-spin concept, and guessed correctly that this was what Bloembergen was secretly driving at. Scovil's and Bloembergen's groups were soon in contact, and they worked out an arrangement. Bell Labs would recognize Bloembergen's priority to the idea and ownership of the basic patent. At the same time, Bell Labs would get to use the patent free for its own purposes. The agreement illustrates a prevailing scientific and business ethic at Bell Labs, which put more priority on being able to use new technologies than on making money specifically on patent rights. It was an approach that fostered a relatively open exchange of ideas.

Bloembergen and Scovil strove to make the three-level maser work, using potassium cyanide. Then, Scovil and George Feher developed still another fillip to the idea. They put together an amplifying system using a crystal of lanthanum ethyl sulfate, a crystal on which Scovil had already done research with Bleaney at Oxford. In the crystal was a small amount of gadolinium, which was to provide the resonance, and a still smaller amount of cerium, which helps relax the gadolinium; this crystal's electron spins relaxed rapidly from the upper to the middle level and led to a working maser.

This episode and many others in maser–laser development illustrate what is perhaps the most critical aspect of a new field—to get it started well enough that the larger scientific community recognizes its importance and potential. After that, the sociology of science asserts its power. Symbiotic, mutually amplifying ideas get traded back and forth among people with a variety of backgrounds and points of view, insuring that the field will develop and grow. The biggest hurdle is the first step, to ensure recognition of its importance and potential.

While activity picked up in the United States, I was still on a sabbatical in which the plan, after all, had certainly not been to look for ways to continue working on the maser. The chance encounter with an old student in Paris and a fascinating new maser material to explore had been surprises that left little choice but to work on it a bit more. After that fruitful winter in Kastler's lab, I left for the next phase of my sabbatical with renewed intent to look into new fields. The maser again receded into the background of my consciousness. It would not, however, remain there.

In early May of 1956, my family and I headed for Japan. Our route was roundabout, taking us through Israel, India, Burma, Thailand, and Hong Kong. Naturally, I dropped in on all the physics laboratories I could. I remember especially well the Weitzmann Institute in Israel, where physicists took turns at night to watch for terrorists, and also the visit with K. S. Krishnan at India's National Physical Laboratory in New Delhi.

I settled down at the University of Tokyo, my aim again to just see what they were doing. As it happened, the faculty there included Koichi Shimoda, who had been a postdoc with me at Columbia and had participated in maser work. It is surprising, perhaps, that my Tokyo encounter with maser physics came not initially from him, however, but from an entirely unexpected direction.

Also on sabbatical there was another Columbia man, a biologist named Francis Ryan. We had known each other pretty well at Columbia. Naturally, we got to talking. He was studying an unusual paper by a British theoretical chemist, Charles Alfred Coulson, devoted to a treatment of microbial population growth. Coulson wanted to describe, quantitatively, the population fluctuations that occur when microbes are both dying and multiplying at the same time. In his paper, Coulson presented and discussed the solutions to an equation that allowed for both the probability of microbe multiplication by division and also a probability of death.

I recognized immediately that this was exactly the kind of mathematical formulation we needed to understand some aspects of the maser, in which photons are both dying (being absorbed) and being born (stimulated into existence) simultaneously, as the result of the presence of other photons. To Coulson's expressions, I knew I had to add another term to account for the spontaneous appearance of photons in a maser—which contrasts with the fissioning of microbial parents—since for microbes there is no chance of spontaneously creating life! But the basic approach, devised for a problem in a field far removed from physics, seemed just what was needed for a precise theory of noise fluctuations and amplification in a maser.

So this second chance encounter put me back on masers in Tokyo. I got together with Shimoda and with a very good applied mathematician there, Hidetosi Takahashi. We three managed to work out a very precise, general theory of fluctuations and of the low-noise quality of masers. Again, I felt it appropriate to publish in a journal of my host country—the *Journal of the Physical Society of Japan*—though this time in English rather than the native language, because my Japanese was definitely inadequate for writing a paper.

With the sabbatical coming to an end, I had things I clearly wanted to do. My encounters and immersions in maser physics convinced me that I should pursue the maser and its applications. The pursuit of higher frequencies was an obvious goal. But a high-frequency maser was not the first thing on my mind. It could wait. Rather, I was set on building, as soon as possible, a good maser amplifier for radio astronomy. I was determined to push hard toward making good astronomical observations with a maser

Figure 10. With my family in 1956, on return from a 15-month sabbatical round-the-world trip, after extended stays in France and Japan. Left to right are Linda, myself, Carla, my wife Frances, Ellen, and in front our youngest, Holly.

at the focus of a microwave antenna. It would amplify, better than any device then available, the faint microwave signals reaching Earth from stars and clouds of gas in distant parts of the Milky Way—and, perhaps, from other galaxies. Equipped with this powerful new technology, I would finally get to work in a field toward which my interests had turned time after time since graduate school.

6
FROM MASER TO LASER

Fired up to put masers to work in radio astronomy, I returned to my old office at Columbia in the fall of 1956. Two excellent graduate students went to work on the project with me, Joe Giordmaine, from Canada, and Lee Alsop. The work also soon drew in Frank Nash, who studied the type of electron resonances that were needed, and Arno Penzias, who came along a little later to work on a more advanced maser.

Today, visitors to Washington, D.C., can still see, on top of one of the Naval Research Laboratory (NRL) buildings, a 50-foot-wide aluminum dish. This radio telescope was the best of its type when my students and I started to work in radio astronomy. Struts can hold an amplifier at the point where the radio waves reflect off the main surface and converge to a focus. With a conventional amplifier, the Navy dish had already made solid contributions to science. Highlights included the discovery of the surprisingly high temperature on Venus. This led to the present recognition that our sister planet suffers from a severe greenhouse effect, making it a hellish place, so hot at the surface that it could melt lead. The NRL antenna was an ideal instrument with which to try our maser. Plus, I had several acquaintances within the Naval Research Lab who could collaborate with us and help us get access to the dish.

A machine shop attached to the Columbia Radiation Lab turned out good hardware for us quickly. Within 20 months of my return to Columbia, we had a maser amplifier on the NRL antenna that was making astronomical measurements. At first, we planned to use a cyanide crystal that Bloembergen at Harvard suggested and that Scovil and others at Bell

Labs had made to work. But before we got very far with that, Chihiro Kikuchi at the University of Michigan showed that ruby could probably work better. Ruby is mostly an aluminum oxide crystal (the mineral called corundum), made pink or red by a small inclusion of chromium. The tough, near-indestructible character of a ruby crystal makes it much safer to work with and less delicate than cyanide! Still more important, the energy structure and the rates at which its electron spins relax from one energy level to another made it attractive for maser use.

We set the amplifier up to work at a 3-centimeter wavelength. This offered several advantages. For one, it permitted a convenient size to the equipment. This wavelength also comes from an energy transition that could be pumped into an inverted population by readily available oscillators. Plus, 3 centimeters seemed a rich wavelength region for radio astronomy. The overall device was about a foot long, wide, and high. The cavity itself was half the size of the wavelength, and it held a ruby roughly the size and shape of dice used in children's games.

Also included was a dewar of liquid helium in which we immersed the resonator and its ruby, and which in turn was surrounded by a dewar of liquid nitrogen. We put a hole through the ruby so liquid helium could penetrate it and keep it cool while an external power source pumped energy in. We needed extreme cold to keep the relaxation rate of the ruby's electrons low. In addition, the cold minimized local heat radiations that might obscure faint signals coming down through the sky from space.

To install the maser in Washington, we worked closely with NRL astronomer Connie Mayer. In building the antenna, the Navy had taken advantage of equipment it could get cheaply and easily. The mount, for instance, was salvaged from a 5-inch Naval gun. It may have been adequate for pointing a cannon, but when the wind blew against the big round dish, it shook so much that it was hard to keep it pointed steadily enough for astronomical work.

Whenever the liquid helium in the maser started running low, we had to climb a ladder into the middle of the dish, take the maser off its mount, and bring it into a room below the antenna for a refill. Usually it went smoothly, but not always. At one point, in the middle of the night, we spilled liquid nitrogen and it ran across the floor to a very large container that the Navy had stored there. The container cracked and out spilled sulfuric acid—considerable time was spent cleaning up the mess. The accident was especially embarrassing because Joe Giordmaine was a Canadian and not supposed to be in that particular Navy room at all.

It took a little while, but eventually the ruby worked very well. One of the first things we did was repeat some studies the Navy astronomers did on a well-known source of microwaves, a nebula called Cygnus A, in the

Swan constellation. This gave us a good check on the performance of our own amplifier and revealed that we were getting an improvement in sensitivity by a factor of more than ten over previous amplifiers. It was wonderfully satisfying to see this device, based on what I had first imagined 6 years earlier on a park bench not so far from the Navy dish, now going to work in a field that had beckoned to me for so long. It was responding to no laboratory or even Earthly signal. A faint trickle of photons that had been born thousands of years earlier in a distant portion of the galaxy entered one end of the maser. From its small ruby burst more powerful surges of radiation, avalanches of photons that represented near-perfect amplification of the signal from Cygnus.

Next, we turned to examining the microwaves from planets, a fairly hot topic at that time. We confirmed the NRL group's previous evidence that Venus is indeed very hot, and we also discovered that the slow-rotating planet is about as hot on its nighttime side as on its sunny side. The explanation seems to lie in the thick blanket of its exceedingly dense atmosphere. We also looked at Jupiter and found it to be colder than dry ice.

Arno Penzias built the next maser to undertake an important, but chancy project—to look for hydrogen gas in intergalactic space. He made a ruby maser amplifier sensitive to 21-centimeter-wavelength radiation, a signal emitted by neutral hydrogen. Only about 6 years earlier, in 1951, astronomers had first used 21-cm radiation to locate concentrations of neutral hydrogen in interstellar space—that is, between stars in our own galaxy. Such measurements produced the first maps of the Milky Way's spiral-arm structure. Yet when we aimed the radio telescope with Arno's maser away from the plane of the Milky Way galaxy into intergalactic space, the instrument recorded nothing. This was a bit disappointing, but not unexpected, and in fact a fine result. A negative answer often can be as informative as a positive one. We learned that what looks like dark empty space between galaxies is, for the most part, exactly that. Of course, Arno eventually did find an important space signal. Only about 4 years later, in 1965, while working with a radio antenna at Bell Labs, he would be the joint discoverer with Robert Wilson of faint microwaves left over from the big bang. That discovery won the two of them the 1978 Nobel Prize in physics, as mentioned earlier.

At about the same time that we were gathering microwave radiation directly from Venus and Jupiter at the Naval Research Laboratory, scientists at Lincoln Laboratory used a maser amplifier to detect radar signals bounced off Venus. The first echoes they got back were weak and hard to interpret. Nonetheless, this was the start of the now-common use of maser amplifiers for the very demanding job of signaling over planetary distances.

We still have in the Townes household the first radio astronomy maser ruby. For our twentieth wedding anniversary, well after the ruby's scientific work was done, I took the crystal to a jeweler in Washington, where I was at the time, and had it turned into a brooch for Frances. The design involves a silver base, which is parabolic—shaped like an antenna—and on it diagrams representing energy levels and also the focuser of the original ammonia maser. The ruby's cooling hole provided a convenient attachment for a silver mounting wire. Maser rubies became popular in the Townes home for a while. Our daughter Linda, when she was about 15, took another ruby, which had turned out to be too highly colored to be useful in masers, and with the help of one of our laboratory machinists made it into a heart-shaped pendant for her mother.

In the late summer of 1957, after the astronomy program was well under way, I felt it was high time I got on toward the original goal that fostered the maser idea: oscillators that worked at wavelengths appreciably shorter than 1 millimeter, beyond what could be produced with standard electronics. For some time, I had thought off and on about this. I had hoped a great idea would pop into my head. Since no such thing spontaneously occurred, I decided I had to take time for concentrated thinking on whatever I could figure out to be the best method. I would have to force a solution, by puzzling through everything I knew about the problems and potential solutions.

To just try things in the lab did not appeal to me. I am generally not interested in performing experiments without a very clear, well-worked-out idea of what will happen. My method of doing science is to figure it out first and know what has to happen. Of course, if the lab shows that nature does not subscribe to the theory I have worked out, then I take a harder look at my theory. Generally, however, I don't try something until I am quite sure it has to work. This seems more efficient than to somewhat blindly try different things in the laboratory. Of course, there are some very good scientists who willingly dive into exploratory experiments without working out much theory or even a clear hypothesis first. Sometimes beautiful, surprising discoveries come that way. It can be a fruitful method, but it is not my method.

The most obvious, straightforward approach to shorten the wavelength at which masers could operate was neither subtle nor at that time very appealing. That would be to build a standard maser, but scaled down to shorter wavelengths, to extend the ideas I had used with the ammonia gas maser. That was in fact the type of idea I had had back in 1951. The problem was that if one simply uses a smaller cavity, fewer molecules flow through it, and they spend less time in there. This would make oscillation more marginal and not allow the shortening of the wavelengths by any large factor over what we already had.

With no ingenious alternatives occurring to me, I plowed through various approaches to see just how far one might go by altering known maser technology for shorter wavelengths. I first re-examined the assumptions that lay behind maser design at the time and the then-common belief that there was no hope of keeping enough atoms in excited states to support any very appreciable short-wavelength amplification. Several good physicists have since told me that they had thought this was the case—the maser idea wouldn't work at very short wavelengths.

The reason for previous pessimism was that the rate of energy radiation from a molecule increases as the fourth power of the frequency, assuming that other characteristics of the molecule remain generally the same. So, a naive estimate would be that to keep electrons or molecules excited in a regime to amplify at a wavelength of, say, 1/10 of a millimeter instead of 1 centimeter, would require an increase in pumping power by many orders of magnitude. Yet, wavelengths at least as short as 1/10 of a millimeter, or even less, would be needed to really get into a new, high-frequency range. Another problem was that for gas molecules or atoms, Doppler effects increasingly broaden the emission spectrum as the frequency goes up. That meant there was less amplification per molecule to drive any specific resonant frequency.

As I played with the variety of possible molecular and atomic transitions, methods of exciting them, and the mathematics governing maser action, what is today well-known suddenly became clear to me: it is just as easy, and probably easier, to go right on down to really short wavelengths—to the short infrared or even optical region—as to simply go down one smaller step at a time. This was a revelation, like stepping through a door into a room I did not suspect existed.

The primary points were, first, that the Doppler effect does indeed increasingly smear out the frequency of response of an atom or molecule as one goes to shorter wavelengths, but there is another, compensating, factor that comes into play: the number of atoms required to amplify a wave by a certain amount is not increased because the higher frequencies make the atoms give up their quanta faster.

Second, while the power needed to keep a certain number of atoms in excited states increases as the frequency increases, the total power required, even to amplify visible light, is in fact not necessarily prohibitive. At microwave frequencies, the lifetime due to spontaneous emission of radiation is in fact very long; the shortening of time an electron spin stays in one state is due to its interaction with the crystal in which it is located, not radiation. For an isolated atom, even if the radiation is billions of times faster at infrared or visible wavelengths, the rate is not necessarily prohibitively fast. Only at still higher frequencies, such as X rays, does the required rate of spontaneous radiation and the necessary exciting power become dauntingly high (in fact, X-ray lasers now exist but they require so much power at these very short wavelengths that part of the system is destroyed or at least badly disturbed every time it is fired).

Not only were there no clear penalties in a leapfrog to very short wavelengths or high frequencies, this process offered very big advantages. The main lure was that in the near infrared and visible region we already had plenty of experience and equipment, such as optical gear. Wavelengths near 0.1 mm and techniques to handle them were relatively unknown. It was time to take a big step.

Still, there remained another major concern: the resonant cavity. To contain enough atoms or molecules, the cavity would have to be large in comparison to the wavelength of the radiation—probably thousands of times greater in dimension. This meant, I feared, that no cavity could be very selective for one and just one frequency. All atomic and molecular transitions occur over a small range of frequency, not at a precise one. This is especially true in a gas maser such as I had in mind, in which the molecules would be moving at various rates and, therefore, would have their frequencies slightly offset from one another because of the Doppler effect.

The great size of the cavity, compared to a single wavelength, meant that many closely spaced, but nonetheless slightly different, wavelengths could find resonant modes. First one would probably dominate, and then another. This would produce a phenomenon called mode jumping. An optical maser jittering from one frequency to another would not be ideal. Nevertheless, it might stay on one frequency for at least a short while. And any oscillator at optical or the slightly longer near-infrared wavelengths could be interesting and useful, so I prepared to proceed.

By great good fortune I got help and another good idea before I went further. After returning from my sabbatical leave abroad, I had accepted a consulting job back at Bell Labs. I had agreed to spend two days each month either visiting there or working on something of interest to them. My assignment was basically to be a friend to the lab, somebody who would be part of the general effort there to interact with the university community and to keep things intellectually stimulating.

After visiting Bell Labs a time or two, one specific suggestion for a visit came from a senior Bell physicist, Albert Clogston. He was later to be the supervisor of one of my former students who became an important laser inventor, Ali Javan, and was already supervising my brother-in-law and former postdoc, Art Schawlow. Al told me that Art had hit a flat spot. Well,

I had always thought very highly of Art, and he was family too. I would have pushed hard to have him stay on at Columbia as a member of the faculty, except that any such plans would have violated the university's antinepotism policy. I was pleased to stop in to see him again at Bell Labs.

Art and I not only discussed his work on superconductivity, I naturally told him what I had been thinking about optical masers. He was extremely interested because he had also been thinking in that very direction. We talked over the cavity problem, and Art came up with the solution. I had been thinking of a rather well-enclosed cavity, with mirrors on the ends and holes in the sides only large enough to provide a way to pump energy into the gas and to kill some of the directions in which the waves might bounce. Art suggested that we just use two plates, two simple mirrors, and leave off the sides altogether. Such arrangements of parallel mirrors were already employed in optics, although for a different purpose and with rather different dimensions; they are called Fabry–Perot interferometers. So, the basic layout was a familiar one. A few years earlier I had even had a student work on a Fabry–Perot system for microwaves. Why I did not come up with that idea for the optical system is hard for me to understand, but I did not, and Art did. Perhaps it had something to do with his familiarity with Fabry–Perots through his earlier thesis work at Toronto.

Art saw that without sides, many oscillating modes that depend on internal reflections would eliminate themselves. Anything hitting the end mirror at an angle would eventually walk itself out of the cavity and disappear, and so would not build up energy. The only modes that could survive and oscillate, then, would be those reflected exactly straight back and forth between the mirrors.

A detailed look and some calculations showed that the size of the mirrors and the distance between them could even be picked so that only one mode or frequency (although of arbitrary polarization) would be likely to oscillate. To be sure, any wavelength that fit an exact number of times between the mirrors could form a resonance in such a cavity, just as a piano string produces not just one pure frequency but also many musical harmonics. In a well-designed system, however, of all the wavelengths that could resonate by reflecting exactly straight back and forth, only one would fall squarely at the transition energy of the maser medium. Other potentially resonant wavelengths would not coincide with the spectral line well enough to reinforce themselves and grow in strength. If one wished to have just one polarization in the amplified waves, meaning that the electric fields oscillated back and forth in one particular direction and not the perpendicular one, this specificity could also be handled with any of a variety of polarizing elements. Mathematically and physically, it was "neat," and all made very good sense.

Art and I agreed to write a paper jointly on optical masers. It seemed clear that we could actually build one, but this would take some time. The maser field was so hot that I felt we should publish our ideas for an optical maser sooner, rather than later, and wait to build one—in contrast to my publishing strategy with the first maser, where publication was delayed about 3 years, until Jim Gordon could make one work as a thesis project. We spent about 9 months working off and on in our spare time to write the paper. We needed particularly to clear up the engineering and some specific details, such as what material to use, and to clarify some of the theory. The paper was done without interrupting the other things we were doing. After having written the microwave spectroscopy book together, we were very compatible coauthors. Art and I would both think about it, work out some idea, talk over the telephone, or get together when I next visited Bell Labs. The finishing touches were actually put in over the telephone and by correspondence while I was in Colorado at a summer study meeting on biophysics.

We were not aware at the time that Bob Dicke, a microwave physicist at Princeton University, was preparing a patent on the use of parallel plates for a submillimeter maser and that Alexander Prokhorov of the Lebedev Institute in Moscow was submitting a paper for publication that involved parallel plates for a maser. Both were microwave physicists, which shows again the field out of which these ideas most easily grew. Their work did not, however, discuss the specific design of such a system to eliminate unwanted modes, perhaps because neither was directed toward very short waves.

Since I had decided rather early that any patent rights associated with the paper should belong to Bell Labs, I felt the whole thing was proprietary. After my initial contact with Art, I did not talk with anyone outside of Bell Labs about it until we gave a copy of the manuscript to Bell's patent department and it had done whatever it considered appropriate. Art could and did talk freely with his colleagues at Bell Labs, getting a number of reactions from them.

Treatment of the idea as proprietary meant that Ali Javan knew nothing about the idea of optical masers until he left Columbia—where he had been a postdoc with me when I first began thinking about them. Once he arrived at Bell Labs and heard about this new line of work, Ali got very interested. As we shall see, he was to be one of the important inventors in the field.

In the summer of 1958, Art and I circulated a preliminary version of the manuscript to some of the people at Bell Labs. Art's boss, Al Clogston, told us he was encountering considerable doubt about whether there could be real resonances in a Fabry–Perot device—that is, between two mirrors with no confining walls. This was another instance in which the difference showed up between the backgrounds of engineers and physicists. To Art and me it was obvious that the idea was sound, but it violated the intuition of many electrical engineers, who were accustomed to closed resonant cavities. Clogston asked Ali Javan to review what we were saying and to advise him as to whether it was really right; Javan's answer was, basically, "Of course it's right." Nevertheless, because some people at Bell Labs were not easily convinced, we decided that I should rewrite and enlarge the discussion of the resonator with a more mathematical treatment.

By August the manuscript was complete. The Bell Labs patent lawyers then told us that they had done their job protecting its ideas, so by late August we sent it around rather widely among colleagues, both in the lab and around the country. We submitted it to the *Physical Review*, which published it in the December 15, 1958 issue.

Before I had discussed the matter with Art Schawlow, a rather peculiar episode in the story of the laser had begun at Columbia. It was also an affair that would prove difficult—although I had no way to suspect so at the time. A graduate student of Polykarp Kusch had talked frequently with me about his thesis. His name was Gordon Gould. Kusch told me that Gould was a bright enough student but somehow just wasn't making progress, and Kusch was afraid that Gould might never get his thesis done. So naturally I tried to encourage him.

Gould was working just a few doors down the hall from my office. He knew about the maser quite early, because it was in the same general part of the building that he used, and he was close to some of my students. Gould was working with a beam of thallium atoms and wanted to study their upper energy states. Following a suggestion that I. I. Rabi had made as a result of some work being done in Europe, Gould's intention was to use an extremely bright thallium lamp to excite the thallium atoms in his beam. In physicists' terms, he was using a thallium spectral line to produce excited thallium atoms so that he could do molecular-beam spectroscopy on them.

Gould had thought about making a maser to amplify microwaves with his thallium beam, and he had come to talk with me about it. He seemed to be very interested in patenting the idea, in any case, whether or not he built one, and asked me a lot of questions about just how to get a patent. I was glad to help. I explained that all you have to do is to have something you really think will work, write it down, have it notarized, and then you have a record of it that will satisfy requirements in case you later wish to apply for a patent. One thing to consider, though, is that the patent has to be applied for within one year after any publication of the idea. He asked me all sorts of detailed questions about the exact procedures.

Gordon Gould never got around to patenting the method for exciting a thallium maser. Some time later—and a little more than a month after the September 19, 1957 date, when I had made a record of my ideas for an optical maser and had them witnessed by my student Joe Giordmaine—I invited Gould to my office and asked him just how much intensity he was getting out of his thallium lamps, supplied by the Edmund Scientific Company. Although we did not want to use thallium, I wanted to know how well the lamps' outputs matched what I calculated I would need to excite an optical maser.

I told him why I was interested, that I was convinced it was possible to make an optical maser. I had not yet talked with Art Schawlow about this, so Bell Labs was not yet involved, and I talked freely, as usual, about any of my work. I recall that after telling Gordon that I believed masers could be made to work this way to generate waves as short as light, he made the comment "I think so too." He was happy to give me what numbers he remembered about the performance of his lamp and in fact, a few days later, came back with updated information, which we discussed at some length. I told him again that I was convinced that an optical maser was practical. It was clear that he was quite interested, but we did not discuss it further, and shortly thereafter I stopped talking with anyone at Columbia because Art and Bell Labs had become involved—and the matter had become proprietary.

About a month later, in mid-November, Gould went to a notary in a neighborhood candy store. In his notebook, he had written down a fairly complete idea for an optical maser, though one without much quantitative theory. He has told people since then that over that weekend after I talked with him he put his ideas into his notebook and had them notarized immediately. His notarized entry was in fact dated substantially later, but his early comment to me may have indicated that he was thinking about optical masers somewhat earlier. As will be seen in the next chapter, his notebook became an exhibit in a court battle with Schawlow and myself over patent rights to the laser. This was to be a conflict that Gould would lose, but it was to be just the opening round in many court cases—and a protracted struggle between Gould and others over rights to a variety of aspects of laser design.

For quite a while, Gould said nothing more to me about masers at optical frequencies, and he left Columbia some time later without finishing his degree to work for a newly formed Long Island company called the Technical Research Group, or TRG. Gould later said that he wanted to work on lasers but that Kusch was too interested in pure science to let him do that. Actually, Kusch was never asked. If Gould had come to me, I am sure that he would have found encouragement. Certainly before Bell Labs became involved, and perhaps even after that, I would have helped him get right to work on optical masers there at Columbia.

The term *laser* arose about this time. It is an amusing history. After we used the word maser, students in my lab kicked around all kinds of corresponding terms. There was *iraser* for infrared amplification by stimulated emission of radiation," *gaser* for "gamma rays," *raser* for "radio," and of course *laser* for "light." In parody, Art Schawlow coined *dasar*, meaning "darkness amplification by stimulated absorption of radiation," which can be translated, simply, into "something black."

Gould was apparently first to actually write down the word *laser*, in his notebook. Initially Art and I were not much in favor of the term *laser*. Maser was the basic device, and it seemed more orderly and systematic to label any variation simply as a kind of maser, such as an optical maser or an infrared maser. Most of our early papers used this terminology. However, laser was of course shorter and easier to say, and as the idea's popularity grew, the device eventually had to have a short name of its own. Its notoriety has today even produced an inverted terminology; I have sometimes seen a maser referred to as a "microwave laser."

When a good idea finally appears, it is very common for a number of other people suddenly to declare that they had been thinking about the same thing all along. They may indeed have been thinking somewhat casually about it and, if the idea appears elsewhere, they may begin publishing on it themselves. Yet if no one else says anything about it, they may never get serious about it themselves.

From the late fall of 1957, when I hushed up about the optical maser—because I felt that it was proprietary Bell Labs information—until our paper began circulating in late 1958, I know of no publications or public documents by anyone about the desirability of optical masers and of no evidence that anyone else was working on such devices during that time. In spite of his stated strong interest, even Gould could not show, in the court patent trial, that he had actually worked toward a laser during that time. Overall, during the months before the release of our paper, the whole field was very quiet. While a number of people say they were interested, an assertion which may well be correct, nobody took it seriously enough to do anything.

I have also observed that it is common, when people try to sort out historic events after the fact, that they may see patterns and may discern motivations that were really not there. Perhaps for that reason it seems natural these days, with so many military uses for the laser, to suppose that the military must have jumped right on it. In fact, some have claimed that the military was pushing in this direction from the beginning, and that is why the laser was developed—but there is documentary evidence of just the reverse. In 1957, I spent part of the summer as a member of a scientific advisory group for the Air Force. It was the second study of such type, designed in the spirit of the first, very successful one of 1945, called "Toward New Horizons." The first study had been led by Theodore von Karman, the famous Caltech aeronautical engineer. The purpose of the second, somewhat as in the earlier study, was to try to predict technologies of importance to the Air Force for the subsequent 25 years. Von Karman was chairman again, but he was old and not really very active in this second study. The meetings were on Cape Cod in Massachusetts, a very pleasant place to take the family, so I agreed to spend a couple of weeks on it, serving on the electronics panel with about ten others.

One of the things I emphasized, of course, was the maser. It was very much on my mind and seemed to have such an important potential. Our report included a recommendation that the Air Force should support further development of masers, for example, as sensitive amplifiers and time standards. We also said that masers should be pushed down in wavelength, at least as far as the mid-infrared region (or about 1/100 of a millimeter), which was my judgment at the time of what was practical. With the military so interested in advanced communications and electronics, and with an unexplored technology in view, maser-related development seemed to me a potentially rewarding investment. We finished up the report in September shortly before my more serious considerations, which convinced me that very short wavelengths could be achieved and which resulted in my notebook entries, as already mentioned.

An additional proposal that came up then and interested me was that space research deserved some attention. The Air Force representatives told us that they had thought about such things too, but they also urged us to please keep anything about space out of the report. They were afraid that Congress would slap them down for fooling around with space "boondoggles" instead of more sensible, serious things.

Well, in October just before our report was to be printed up, the Soviets launched *Sputnik I.* All of a sudden, Congress wanted the Air Force to be in space; everybody did. Nobody with any brains would ignore space. Except, of course, by instruction, our report did exactly that. Since it would have been terribly embarrassing for the Air Force to publish a report on its future directions for the next 25 years and *not* say anything about space, the Air Force immediately put the report on hold.

That was not the end of it, however. The Air Force called most of the group together again the next summer, in 1958, to revise and polish up the report. The document then, of course, had something to say about the military potential of space activity and research.

More important for this story, I was not part of the 1958 group (my "vacation" that year with the family was in Colorado, with the aforementioned biophysics group, during which I was also finishing up the paper on optical masers with Art Schawlow). The Air Force study group that did meet, however, decided to leave out the section about pushing masers to shorter wavelengths, such as the mid-infrared region. The paper that Art and I were then writing only became generally available in August of 1958, and presumably most members of the Air Force study group were still ignorant of it. As already noted, there had been almost no talk about optical or infrared masers during the prior 9 months, so the electronics advisory group for the Air Force, rather than pushing the field, actually decided to drop my suggestion of the year before that it might become important.

The Air Force committee's members either felt that masers could not be pushed to wavelengths as short as the infrared region or that they would have too little military relevance to merit Air Force interest. For the report, they left in only the part that I had drafted about masers for the microwave region. Without me there to push the idea of working with much shorter wavelengths, all enthusiasm for it left. Perhaps the other members just thought it was a pet and impractical project of mine. Air Force representatives had, of course, seen the section on infrared masers in our draft from the year before, but none of them fretted over its disappearance from the final version. Today, some may continue to believe that the military was pushing the laser hard during those formative years, but the real evidence is in just the opposite direction.

At the same time, it is true that some individual military representatives followed and understood our work. In late 1956 or early 1957, Bill Otting, an employee of the Air Force Office of Scientific Research, stopped by to see me. He had a master's degree in physics, was the Air Force liaison to the Columbia Radiation Laboratory, and enjoyed seeing what we were doing. He urged me, on his own, to write an article about pushing masers into the infrared. Well, I told him that shorter wavelength masers were just the thing I wanted, too, but so far, I hadn't had any great ideas of how to do it. If I got the right idea, I told him, I would gladly write an article, but at the time I did not have one. He asked who else might give it a try, and I suggested my postdoc Ali Javan, who was quite outstanding, particularly on the theoretical side. But Javan decided against it, too, so the paper did not get written. This was all before the 1958 Air Force summer committee, which in the end did not say anything about shorter wavelength masers either.

After our paper on optical masers began to circulate, the lack of official military support for shorter wavelength masers did not last long. In early

1959, Irving Rowe, a representative in New York City of the Office of Naval Research (ONR), asked me if it would be a good idea to sponsor a conference on masers, which by then had become a hot topic. He could get the ONR's Electronics and Physics branches to provide money for it, and would I be chairman? I agreed. The ONR gave Columbia a grant for the meeting, including enough money for a secretary and an assistant.

I put together an 11-man steering committee. It included Rowe, Nico Bloembergen of Harvard, Bob Dicke of Princeton University, Tony Siegman of Stanford, George Birnbaum of the Hughes Research Lab, Charlie Kittel of the University of California at Berkeley, Woody Strandberg of MIT, and Rudi Kompfner of Bell Labs. I tried to include both physicists and engineers, ranging from those good in theory to those best at understanding practical problems. I decided against a very large meeting, and settled on a target of 100 key people. A lot of the organizing work could, fortunately, be carried out by my students, including Pat Thaddeus, Harold Lecar, Joe Giordmaine, Bill Rose, Frank Nash, Isaac Abella, and Herman Cummins.

Early in the summer of 1959 and before the planned meeting had occurred, I agreed to take a leave from Columbia to become director of research and vice-president of the Institute for Defense Analysis (IDA) in Washington, D.C. This was a time when post-*Sputnik* worry still ran high that the United States, including its military, was falling behind the Soviet Union in important science and technology. IDA was a nonprofit advisory body to the Pentagon, managed largely by a group of university presidents, including the president of Columbia, my own university. I have always felt that scientists should provide public service from time to time, and so I agreed to go to Washington to help monitor American scientific work and advise the Pentagon and other parts of government on needed development of technology. I planned to be there for 2 years and then get back to an academic setting.

During my stay in Washington, I went to Columbia regularly, on Saturdays, to check in on my students and to oversee planning for the meeting on maser physics. The 1959 meeting represented the formal birth of maser and related physics as a distinct subdiscipline. For a title, our Committee settled on "The Conference on Quantum Electronics—Resonance Phenomena." It seemed descriptive yet broad enough to cover all the then-pertinent topics, and this was the origin of the term *quantum electronics*.

We met September 14 to 16 at the Shawanga Lodge, a resort hotel in New York State's Catskills. Rowe declared in his opening remarks that quantum electronics "is actually producing a revolution in microwave techniques." He pointedly added, "Those of you who are not familiar with the Office of Naval Research may wonder why the Navy should spend its money on a scientific conference of this type." He answered by noting the use of a maser on the Navy's radio telescope, as well as the Naval Observatory's role as official national timekeeper, including use of an atomic clock. The ONR's interests, he explained, "are not devoted primarily to immediate practical applications . . . we are interested primarily in encouraging basic scientific research, with the emphasis on providing a better understanding of the fundamental processes of nature." These were fine and true sentiments. The meeting exemplified the vital role played by the Pentagon in supporting basic research during the late 1940s and the 1950s. The National Science Foundation (NSF) and other nonmilitary sources of government support were not as prominent then as they are today.

The meeting largely explored the confluence of the fields of spectroscopy and electronics, with masers the dominant but not sole topic. In addition to discussions of the theory and practical aspects of masers, there were papers on atomic clocks, on nonmaser amplifiers, and on the basic physics of atomic- and molecular-energy transitions. The 60-plus papers served as catalysts to the debates, conversations, and shop talk that filled the days and evenings. It was a hardworking conference. Even though it was held at a resort, few participants brought spouses.

This was the first of what was to become a long series of international meetings on "Quantum Electronics." A recent one that I attended was swarming with more than 7,000 people, as compared with my plan for 100 at this first meeting. Also, now there are always large and impressive exhibits of commercial products—from publications to laboratory and medical instruments, to gadgets and manufacturing equipment.

Jim Gordon opened the 1959 conference proceedings with a review of molecular-beam masers, including reference to the work in Russia by Basov and Prokhorov, who were in the audience and on their first visit to the United States. After the conference, they visited our lab at Columbia and Frances and I were pleased to have them over at our house.

On the last day of the meeting, Art Schawlow reviewed our conclusions on optical masers. Javan discussed gaseous optical masers, including his idea for "collisions of the second kind." Gordon Gould had been invited to the conference. He did not present a paper, but as I expected he had some comments about Javan's talk. In late 1958 and in 1959, after Gould had left Columbia for TRG and had seen our paper, he came to Columbia from time to time and talked with me. One thing I remembered him saying was that he had a new idea for exciting atoms to produce an optical maser, using collisions "of the second kind." He wanted to take advantage of a well-known physics phenomenon. If an atom with electrons in an excited state collides with another atom in a lower state, sometimes the energy may transfer from the excited atom to the other one; this can happen even

Figure 11. At the first international conference on quantum electronics, 1959. Left to right are James Gordon, Nikolai Basov, Herbert Zeiger, Alexander Prokhorov, and Charles Townes.

if they are atoms of different elements, so long as the amount of energy exchange fits a natural transition in the atom on the receiving end. It is not a transfer of kinetic energy, but one of electronic, or internal, energy.

Entirely independently, Ali Javan at Bell Labs had already talked with me about the same idea. I had encouraged Ali to pursue and publish it. Javan never considered using the ruby or any other solid system that Art and I had thought about. He has told an interviewer simply that "I always looked to gaseous media . . . I don't do solids. I prefer the simple interactions of single atoms or single molecules."

After Gould told me of his interests in collisions of the second kind as an energy source, I found myself entrusted with two nearly identical confidences—and, of course, encouraged him to pursue and publish it. I could not reveal either person's interest to the other. Javan worked it out much more completely than Gould did and published in *The Physical Review*, in 1959, but I was amused that both had independently hit on the same idea and that it was one Art Schawlow and I had not thought about. This kind of keeping of confidence about fast-growing laser ideas occurred more than once. As semiconducting lasers came along I was told confidentially, by three different groups, two of which were in large commercial companies, of their ideas for such lasers. None was aware of any other's work—I kept mum. And there were other awkward matters about who had or should have information.

In December 1958, the TRG company, where Gould had gone to work, submitted a proposal to the Pentagon's Advanced Research Projects Agency (ARPA) to work toward a laser. The ARPA had been set up so that the United States could vigorously compete with the Soviet Union after the shock of their sending up *Sputnik*. ARPA was given plenty of money to support new ideas and technical development and was dispensing money very liberally, at that time, on new technical proposals. TRG needed an infusion of support and hoped its proposal would bring in a good contract. The U.S. Department of Defense asked me to review the proposal, which stated as its purpose "To Study the Properties of Laser Devices." It was the first time that I knew TRG was seriously in the game. Larry Goldmuntz, president of TRG, later told me he had to sit down and work with Gordon Gould to get the proposal written. It outlined a project that would support seven of TRG's staff members. It was a very long document, as such proposals for the military generally are. I said that I didn't really have time to review it, but somebody at the Pentagon called again. He told me TRG insisted that it not be reviewed by anybody but me. Presumably the company did not trust their proprietary information with anybody else and knew that I was fairly familiar with it, anyway, and would look on the field favorably. So, with this insistence, I accepted.

The TRG proposal discussed the theory of lasers, resonant modes of Fabry–Perots, communications and laser radar, and transmission of power to a satellite, though not laser weapons. Most of their proposal indicated what might be done if laser power were in the range of a few watts, but even 100 kilowatts was mentioned as a possibility for ultraviolet radiation.

Of course, nobody had made any lasers at all to that point. The ones we discussed in our publications would have outputs of a few milliwatts. What TRG discussed in detail they estimated would produce 100 milliwatts. But TRG mentioned, in broad terms, its hopes of producing lasers with outputs as high as 10 watts. Today, we can see that TRG's open-ended mention of many watts was right on target, though hard to justify at the time. Some giant lasers now put out beams with as much as a megawatt (1 million watts) of continuous-wave power, though not at the ultraviolet wavelengths proposed by TRG.

So, I looked it over and said, yes, this is valuable work and the field ought to be developed. I couldn't be sure that kilowatts could be produced, but a constant output of hundreds of watts seemed quite possible. The proposal provided for no specific results but was simply to study "properties of lasers." While I was glad to recommend funding, and ARPA was enthusiastic enough to allot TRG $1 million (the company had only asked for $300,000), I was also a little annoyed with TRG's proposal. It had copied so much out of the paper that Art and I had written but gave us no acknowledgment or reference. The next time I talked with Gould, I recall asking, "Well, Gordon, haven't you seen our paper?" He said yes—and indeed he had, as an entry in his own notebook was later to show.

TRG kept struggling to get going and I wanted them to succeed, though their laser work was not progressing especially well. So, when it came time for the 1959 conference, I asked Art if, in his general talk on optical masers, he might say something good about Gould's work. If it did fit in, I felt it might give Gould and TRG a little boost. And Art did manage to provide some positive comments.

At the meeting, Gould took part in the general discussion following Art's talk, and mentioned there the possibility of pulsing lasers which, he estimated, might produce peak powers as great as 1 megawatt. This was a daring prediction to make before anybody had made any laser at all, but it has turned out to be quite correct.

Also listening with interest was Theodore Maiman from Hughes Research Laboratories. Ted had gone to Hughes to join the atomic physics group, organized by Harold Lyons, and was establishing a reputation as a good practical physicist and experimentalist. He had been working with ruby masers and, earlier in the meeting, reviewed some of his work on those microwave devices. Ted says that he was also already thinking about using ruby for an optical maser or laser.

Art's talk dealt mostly with prospects for a potassium-vapor gas laser pumped with a potassium lamp and with Ali Javan's subsequent proposal for a helium–neon gas laser. But Ted listened perhaps most closely to Art's rundown on the possibility of a solid-state ruby laser. Art concluded that the energy levels of pink ruby were not suitable for a laser. The difficulty was that the level to be used as the lower state was also the ground state, which would seem hard to keep empty enough to be the bottom of an inverted energy scheme. Instead, he suggested using darker ruby, where there were useful energy levels without this problem. Art commented that, "It may well be that more suitable solid materials can be found, and we are looking for them."

Ted said later that he felt Art was far too pessimistic and that he himself left the meeting intent on building a solid-state pink-ruby laser. In any case, in subsequent months Ted made striking measurements on ruby, indicating that the lowest energy level could be at least partially emptied by excitation with intense light. He then pushed on toward still brighter excitation sources. On May 16, 1960 at the Hughes Laboratory in Culver City, California, Ted fired a General Electric flash lamp that he had wrapped around a ruby crystal about 1 centimeter across and 1.5 centimeters long; it produced the first operating laser. The evidence that it worked was somewhat indirect. The Hughes group did not set it up to shine a spot of light on the wall or anything like that. During early discussions of their success, I asked a senior Hughes scientist repeatedly if any flash of a beam had been seen. That seemed the obvious test, but the answer was that no such flash had been seen, and this created some room for doubt about just what was obtained. However, the instruments had shown a substantial change in the fluorescence spectrum of the ruby; it became much narrower, a rather clear sign that emission had been stimulated, and the radiation had peaked at just the wavelengths that were most favored for such action. Excited by a flash lamp, it was in fact a pulsed laser, though the proof seemed initially to be only that more monochromatic radiation was produced. A short while later, both the Hughes group and Art Schawlow at Bell Labs independently demonstrated powerful flashes of directed light, which made spots on the wall—clear intuitive proofs that a laser is working.

The Hughes Research Laboratory announced this first laser to the public at a press conference July 7, 1960, in New York. The next morning's *New York Times* had it on the front page, with the sedate headline "Light Amplification Claimed by Scientist," and made sober reference in the story to earlier maser work. Some other accounts were not so reserved, with a few hinting broadly that Hughes Labs had invented a killer-ray gun with a future on battlefields to come. Ted said later that he was surrounded by reporters following his formal presentation. He had put weaponry far down his list of possible uses, well below its scientific applications and use as a measuring device. The reporters pressed him to be more enthusiastic on the weapons angle. Exasperated, one writer asked Ted if he could rule out the laser as a weapon. Ted said no, he could not rule it out—the laser death-ray story was born.

Hughes distributed press releases far and wide. One of the Hughes photos that the newspapers particularly loved showed Ted's face, including a pair of protective goggles, nearly filling the entire frame. He was holding the laser right in front of his nose. The picture shows a ruby crystal, about 1 centimeter across and several centimeters in length. Such a relatively long, thin proportion was in fact a good and common design in later ruby lasers. For many scientists, such proportions became fixed as the archetypal ruby laser. However, it was the wrong ruby. The real ruby was short and chunky. *Physical Review Letters* had been besieged by maser reports, and its editors, to their later chagrin, turned down Ted's paper without having it refereed. They mistakenly thought it was a follow-on of something he had already published—just another maser study. So Ted turned to *Nature*, which accepted a brief description of his historic accomplishment and published it August 6. That short paper, to my puzzlement and probably that of many others, described the heart of the laser as "a ruby crystal of one cm dimensions." There was no question at that time that he had beaten the rest of the world to the first laser, but the dimensions of what that first published picture showed did not match the dimensions of what Ted and his associates had reported.

Recently I learned that Ted, now living in Santa Barbara, has a simple explanation. When the public relations people at Hughes learned what Maiman had done, they sent a photographer around to his laboratory. Ted picked up his first laser with its stubby crystal. The photographer thought it did not look nearly dramatic enough and spotted nearby a longer and much handsomer ruby crystal. The reason for the extra crystal was simple. Ted had ordered three flash lamps from General Electric. He decided to use the smallest one. Yet for backups he was prepared for other prototypes, using larger crystals and bigger lamps. Ted went along with the photographer because that backup did in fact work fine. As a result, the second or third laser, not the first, got its portrait published around the world.

As a practical matter, none of the details were critical, of course. Many excellent people were pushing hard toward lasers. And while the ruby laser was a starter, and continued to be important in experiments with intense light, the variety of lasers and laser techniques exploded. Lasers working in gases and solids, even in liquids (and Art Schawlow's Jell-O laser mentioned earlier) followed rapidly, with the myriad applications that suffuse modern technological society. Two other kinds of solid-state systems were the next to work, put together at the IBM labs by one of my former students, Mirek Stevenson, and one of Nico Bloembergen's former students, Peter Sorokin. Then Ali Javan's helium–neon laser very soon came along, built by him, by Bill Bennett, another student who had gone to Bell Labs from Columbia's Radiation Lab, and by an optics man at Bell Labs, Don Herriott. The helium–neon laser, with some modifications, was in fact for a long time the basis for the world's most common lasers. More recently, with the advent of tiny semiconductor lasers that have been produced by the millions and which had their origins in solid-state physics, the helium–neon laser has been outnumbered.

Our group at Columbia Radiation Lab had been working on lasers while I was at the IDA in Washington, D.C. Since it was to be a student thesis, and in spite of enthusiasm for the work, we moved along at something like normal thesis speed—too slowly. After the first laser came along from Hughes, my students, Herman Cummins and Isaac Abella, wrote their theses on excitation phenomena in cesium vapor and some of the earliest work on two-photon absorption, which the laser intensity made possible. A bit later, one of the types of lasers described carefully in the paper by Art Schawlow and myself was made to work by a group at TRG, but it was only a demonstration and never a very useful type.

It is notable that almost all lasers were first built in industrial labs. Immediately after our first publication on optical masers, or lasers, it became a very hot field. Part of the reason for industry's success is that once its importance becomes apparent, industrial laboratories can put more resources and concentrated effort into a problem then can normally be done at a university. When the goal is clear, industry can indeed be effective. Yet the first lasers, though built in industrial laboratories, were invented and built by young scientists recently hired after their university research in the field of microwave and radio spectroscopy—students of Willis Lamb, Polycarp Kusch, and myself, who were together in the Columbia Radiation Lab, and of Nico Bloembergen at Harvard. The whole field clearly grew out of this type of physics thinking and experience.

For me, work on the laser itself was coming to an end. From that point, I would employ it extensively for scientific experiments but leave to others the continued, and remarkable, expansion of laser technologies. What I could not know was that patent problems and court cases would cause me for some years to go over, and over, and over, that phase of my career in which the laser first came to exist.

7
THE PATENT GAME

The credit for invention, and even the meaning of invention, is a slippery thing. Take, for example, phase stability and the synchrotron, a type of circular particle accelerator used for high-energy physics.

Phase stability employs a clever method for adjusting the timing of the acceleration of particles to compensate for the growing masses of particles as they approach the speed of light. It overcomes handicaps that prevented particle accelerators of the time from working at relativistic energies. Ask any physicist who invented the high-energy synchrotron, and the likely answer will be the late Edwin McMillan. Ed, who won a Nobel Prize in chemistry in 1951 with Glenn Seaborg for the synthesis of plutonium, was a physicist at the University of California, Berkeley. After the death of Ernest O. Lawrence, Ed became director of the Lawrence Radiation Laboratory (now the Lawrence Berkeley National Laboratory). He was the first to push synchrotrons beyond the energy range of 100 million electron volts. This stimulated an explosive growth of the field and was a major reason for the laboratory's continued excellence in high-energy physics for many years after Lawrence's death.

As with many "original" ideas, some aspects of Ed's phase stability idea, which allowed these high energies, had already occurred. A good paper sent in for publication in March, 1945, by the Russian scientist Vladimir Veksler surfaced that outlined its theoretical principles quite well. No one had paid much attention to it, and Ed never saw that paper before he sent in his own paper for publication in September, 1945. Shortly after Ed published the idea, physicists at General Electric were the first to try it out on a small high-energy machine which was already available to them.

Asked about these competing claims, Ed once commented to me, "All I can say is that I was the last one who had to invent these machines." There is wisdom in Ed's statement. It is a fine thing to have an idea, and still better to write it up so that other people might use it. It is even better to be the one who not only thinks of something independently, but who makes the invention real in such a way that it is unnecessary and in fact impossible for anybody to invent it again.

Arguments over invention, and multiple claims for the credit, would be mainly fuel for after-dinner arguments about prestige were it not for patent law. Patents are critical means for determining who gets the money from inventions, and who has to pay to use them. In the courts, patents are very serious business; Ed's insightful joke would not go far with a patent judge. In important cases, legalistic decisions determining who invented what can determine the fates of business empires and consume many long months of maddeningly convoluted legal argument in court.

Scientific discovery is generally thought by many to be similar to invention. Indeed there are similarities and sometimes the two are closely connected. But there are also striking differences. As will be seen in the case of masers and lasers, patents depend on legal rules in which technicalities—such as minor variations on the main idea or whether an idea was officially public—are critical to granting patent ownership. In addition, the importance of lawyers, juries, and financing leads to special peculiarities that are strange to science. Furthermore, what is patentable is a device, not a new principle or a discovery about nature. New discoveries by scientists about nature or its basic laws are not new from a patent point of view; they have already been in existence. The scientist works for the excitement and satisfaction of new understanding; the reward is in part wonderful new insights into nature and, in part, the recognition of success by oneself and by peers. Scientific progress arises from discovery of these new insights or of principles that are new, in the sense that humans have not previously recognized them even if they have been natural law since the beginning. But a person oriented primarily toward invention must aim for new devices not occurring in nature that can be put to practical use. The reward may include great satisfaction in providing a useful device, or it may be primarily monetary. Obtaining a patent in the United States involves the rules of patent law, set up more than 200 years ago by thoughtful people but necessarily with many categorical distinctions and specific rules that may or may not apply neatly to specific cases.

The scientist's instinct is to speak freely with colleagues and share ideas; concern about patents, however, discourages openness because it can obscure precisely who invented what, a determination that is the main point and legal requirement of a patent. Cooperation and free discourse are hallmarks of civility and almost inevitably enhance collective creativity; but when financial rewards from patents are large, they may set colleagues on courses of secrecy and nasty collisions in court. There are thus incompatibilities between the optimum culture of science and the labyrinths of patent law, in addition to some broad overlaps.

During the war, when I was at Bell Labs, I was carefully instructed on the practical importance of recording technical developments. I was to be alert to anything that might generate a patent. The point there was not so much to make royalty money but to see that the company was free to use the latest technology. Since any patent resulting from Bell Labs research was company property, the process did not discourage openness and collaboration within the company. While there, my work produced about a dozen patents, including one for the molecular or "atomic" clock. My first brush with the real passions and expensive entanglements that arise from patent disputes came after Bell Labs.

When I moved to Columbia University, I had occasion to pass time in the Faculty Club with Major Edwin H. Armstrong. He was a professor of electrical engineering and a remarkable inventor. For example, his inventions included regenerative circuits, superregenerative amplifiers, superheterodyne detection, and wide-band FM radio. He was very creative and greatly interested in patenting whatever he could. Already by the age of 25, his patents had made him a millionaire. Protecting and defending his patents became a consuming preoccupation.

When we met, usually over lunch, a frequent topic was his bitterness over the difficult struggle to defend his patent on frequency modulation (the technique underlying FM radio). A number of companies, including RCA, refused to pay the royalties that Major Armstrong felt were his due. And he had already spent a large part of his considerable wealth fighting for these royalties.

One day, I was shocked to learn that Major Armstrong had killed himself. He jumped out of a thirteenth-story window. No one knows the exact reason; I understand he had some personal problems, but patent problems and their costs, which brought him close to bankruptcy, were obviously a major source of his anguish. After his death, his estate's lawyer, Dana Raymond, pursued the FM case vigorously and won. The Armstrong heirs received about $10 million and his estate was restored.

One thing about pursuing patents that must be emphasized is the critical importance of money. Patent cases can be dragged out for such a long time that anybody who does not have staying power can get run over. It takes money to get a good lawyer, and a good lawyer in control, perhaps even more than the merits of the case, makes a whale of a difference. That's what made the difference for the Armstrong heirs. It is impossible for me to imagine suicide over money or a patent dispute, but the Armstrong episode helped impress on me the hazards of letting patents, royalties, and worries over lawyers intrude too far into a scientific career.

Of course I always welcomed extra income for my family. Patent money is not tainted, and I don't wear a hair shirt. When I have to deal with patents, I try to be careful to guard my interests. But maximizing income is not paramount. Science and other values come first. Instinctively, it does not feel right to me to do something just for the money involved, and I don't want to waste time that way. I never imagined that any of my Bell Labs patents would be worth a great deal of money—but the maser was different. When I told a fellow physics professor at Columbia about it he immediately said: "Hey, you'll make a million on that!" I said "Maybe," and laughed.

Columbia in the 1950s had a rather freewheeling approach to patents, with no clear-cut policy. The agreement we signed in order to use government money for research held that any patent resulting from it, such as my own, was to be taken out by the university, with rights of free use assigned to the government. There was a kind of standing committee to review such patent possibilities, but it was not very active. When I told the chairman of the patent committee that the maser ought to be patented, he told me the university actually had no clearly set procedure for such matters. If I wanted to patent the maser, the committee decided, I should just do it myself. It was clear the device should be patented but, again, I did not want to get too embroiled in any patent involving a potential briarpatch of legal conflicts, even though it might generate a lot of money.

Fortunately, there was (and is) an outfit designed to take over patent responsibilities, which suited me perfectly. This is the Research Corporation. It was set up in 1912 by a chemist and inventor, Frederick G. Cottrell. He was a man of enormous energy, ingenuity, and principle. Cottrell grew up in Oakland, California, went to the University of California at Berkeley, at age 16, where he got his degree in just 3 years. He later became a member of the National Academy of Sciences, in 1939.

In 1907, Cottrell had attached a device of his invention to the smokestacks of a lead smelter on the Carquinez Strait, north of Berkeley. The smelter's owners had been facing ruinous lawsuits from surrounding farmers infuriated that the plant's fumes were wrecking crops and even corroding household window screens. Cottrell had built the first successful electrostatic precipitator; it used a powerful electric voltage to remove mists and particles of pollutants from the smelter's stacks. This patent promised to make him a fortune, but in a display of public spirit Cottrell turned over the management of the patent and its revenues to a new kind of corporation, established in cooperation with the Smithsonian Institution in Washington, D.C. Cottrell wanted the money to help science produce useful products. As he told the American Chemical Society at the time, "There is a certain amount of intellectual by-products going to waste at present in our colleges and technical laboratories all over the country (that) dies right there because the men . . . do not want to dip into the business side of technology."

These days, when so many governmental agencies, as well as many private foundations, provide multiple sources for research grants, it is important to know how hard it was early in the twentieth century for most investigators to get the wherewithal to pursue novel research. The Research Corporation played an important role in nourishing American science, particularly in the years leading up to World War II. Ernest Lawrence, for one, depended in part on the Research Corporation's grants to develop his early cyclotrons in Berkeley. I. I. Rabi at Columbia was awarded Research Corporation money for his molecular beam work. The corporation continues today, from its headquarters in Arizona, to devote money to science, mainly through support of researchers at smaller colleges.

At any rate, it was an organization I admired. It could also take patent problems out of my hands so they would not be too much of a nuisance. I asked Research Corporation officers to take care of the legal work on the maser patent so that I had only to worry over the technical information. I stipulated that any income from the patent should go first to pay off the corporation's legal expenses, which were likely to be large. After that, I was to get 75 percent of the income, until I got $25,000 per year. Why $25,000? Well, that was about three times my annual salary at that time, and I thought I didn't really need more than that. It was plenty of money. After that, the corporation was to get 75 percent, I would get 20 percent, and 5 percent was to go to Jim Gordon. Although Jim did not have a piece of the patent, he had done so much to get the first maser working that I felt he should get a share of the downstream royalties. In addition, $1,000 was to go to Herb Zeiger, the postdoc who helped build the maser, and $1,000 to Tien Chuan Wang, the Chinese immigrant who had assisted with the second maser and helped demonstrate the beautiful performance that masers could deliver.

United States law requires that for patent protection, one must apply for a patent within a year of the first publication of an idea covering a patentable device or system. The first maser publication was in June 1954. I had published before worrying about a patent, but with the law in mind we just made the deadline, filing for a patent almost a year later. The first lawyer I worked with at the Research Corporation was Harold Stowell, a somewhat elderly man. Stowell and I had to interact a good deal to properly formulate the maser patent, and we kept up a correspondence on it even while I was on sabbatical in Europe and Japan. I was particularly careful to make the patent language as broad as possible, because the new idea opened up a field much broader than just ammonia masers operating at microwave frequencies. I wanted to patent a technique for amplifying electromagnetic waves in general. The basic method clearly would work at wavelengths other than microwaves and with media other than ammonia. We made certain to craft language that covered the inevitably wide variety of applications—across the frequency spectrum and with an unrestricted range of materials.

The patent application involved some very specific as well as very broad claims—enough, I felt, to cover the field. Getting the actual patent granted, however, took some years more. Things seemed to be going slowly; I was eager to get it done, and I told the Research Corporation's lawyers as much. That is when I got some hard-nosed business instruction in patent law. They explained that it is seldom wise to move fast on finishing up a patent.

While it may be important in science, as well as in patent law, to work quickly to establish one's ideas, a patent usually makes more money if the date that it takes effect is as late as possible. It is a matter of money and, with most really new devices and ideas, the biggest revenues come many years or even decades after their invention. Patents expire after 17 years. Under the laws then applicable, the more one delayed the start of a patent's life, the longer it would survive into those 17 years when the invention could make big money. In fact, with many inventions, it is not the original patent that makes the most money but the add-ons, the little patentable fillips and alterations that come along later, piggy-backing on the original invention and taking advantage of its most fruitful years. While the maser idea was published in 1954, and the patent application was made not quite a year later, the patent itself did not go into effect until 1959, just before the first laser was built. The lawyers at the Research Corporation were dead right. If we had maneuvered to delay the final patent even longer, for another ten years or so, a great deal more money would have come in.

By contrast with the Research Corporation and the maser, the patent for the laser was handled by Bell Labs. Of course, the maser patent was written broadly enough to cover a wide range of wavelengths, including those of light, which made it the primary patent. Although there were enough new things in the laser to qualify it for a separate patent, it nonetheless was to be a subsidiary patent to that on the maser.

A number of ethical quandaries had to be faced before I decided to assign the laser patent to Bell Labs. It was not entirely obvious where the patent ownership should reside. The organization paying for my work should be the legal owner of my part of the patent. If that was Columbia University, then my part of the patent would eventually be mine, as was the maser. In fact, I had started work on the laser in my office at Columbia. Only later did I talk at Bell Labs with Art Schawlow, who added important ideas. Should ownership of the patent be shared jointly by Columbia and Bell Labs? Given to one or the other? My official working obligation to Columbia was 40 hours per week, but of course I worked more than that. Furthermore I was entitled, as was the custom, to spend one day a week consulting, and I had an agreement to consult for Bell Labs on a freewheeling basis without much attention to location or precisely counted hours. Which hours belonged to whom was no clear-cut thing. Settling the matter was up to me. In the end, I felt it could legally go either way and, since the maser patent had the whole field covered anyhow, it seemed rather self-serving for me to try to own part of the laser patent when Art Schawlow at Bell Labs was clearly an important contributor. So I decided to make it simple and assign it all to Bell Labs.

As for the patent procedure itself, I asked Art, because he was working full time there, to take a copy of our paper discussing the "optical maser," as we called it then, to the Bell Labs lawyers and have them work up an application. I suggested to Art that he approach Arthur J. Torsiglieri, a top patent lawyer for Bell Labs. He prepared the patent on transistors, and Bill Shockley, who was one of its inventors, had told me he was very good. However, the job was assigned to a more junior lawyer. To my amazement, Art came back and said the patent lawyer didn't seem very interested in patenting it! "Why not?," I asked. Well, Bell Labs and its parent corporation, AT&T, were in the communications business, and the lawyer didn't think that light waves had much to do with communication. Memories of the patent lawyers, Art Schawlow, and myself differ a bit on details, but what I remember includes Art's report to me that if we wanted it patented we could go ahead and do it ourselves. I felt that this was just a lawyer's misunderstanding as a result of inadequate technical imagination, and that it would be unfair to Bell Labs if we did not urge them along. Perhaps the lawyer saw the whole thing as just a cute specialized scientific idea. I suggested to Art that he go back and talk with them some more. This time, he reported that if we could show applications to communications, they would patent it. One of the lawyers even told Art a story about how Alexander Graham Bell, founder and patron saint of the lab, had tried using light for communication. By a most strange coincidence, Bell had worked on his idea in a building next to Franklin Park in Washington, D.C. the very place where the maser idea was to occur to me many years later. But Bell's ideas were unsuccessful. His failure perhaps left a bias in their minds against sending messages with light. Nevertheless, the lawyer told

Art that if we could show applications to communications, Bell Labs would have a clearer case for patenting it.

In the end, we passed muster with the lab's patent lawyers by describing possible applications for sending messages on beams of light, which was in fact easy to do. To emphasize this aspect, the patent was entitled "Optical Masers and Communication."

I had become a bit fed up with the Bell Labs lawyers' lack of enthusiasm for the laser patent, and I was busy with other things. I had seen to it that Bell Labs got the patent, felt they had experienced lawyers who ought to be able to do the job well, and so paid little more attention to it. This was a mistake on my part. The lawyers could not really be expected to understand very completely the technical potential and ramifications of such an entirely new field. At that point, scientists didn't either. I had worked assiduously on the maser patent to see that it covered everything appropriately, and it proved in the long run to do so. But as a result of my and Art's inattention, only the basic idea in our optical maser or laser manuscript was adequately covered in the laser patent application. Several extensions and variations that were mentioned in our manuscript had not been covered well at all, so claims on them were made later by other people. I could probably have saved Bell Labs and myself a lot of later trouble if I had paid more attention to the patent draft.

A related quandary arose during this time. At the very same time that I was helping the Research Corporation attorney elaborate the basic maser patent, and in that case paying careful attention, the Bell Labs lawyers were framing the subsidiary laser patent. The interests of those two groups of lawyers were in some ways quite opposite, for to broaden one patent might limit the legal territory left to the other. The idea of amplification by stimulated emission of radiation was clearly fundamental, and it had many ramifications, but how could they all be fairly and adequately covered by the two competing organizations? The Research Corporation and I wanted the basic maser patent to be as broad as possible. But I felt I could not even hint to their lawyer that I was working on an optical version, which was Bell Labs' property. So there I was, involved in two patents being filed by different parties and keeping each secret from the other. One decision I made on my own was to leave out any specific mention of light waves in the maser patent. Whereas the maser patent covered the entire electromagnetic spectrum, I felt that explicit description of possible visible light applications could be unfair to Bell Labs' interests.

This was not the only conundrum in which I was to find myself. At one point a bit later, I got caught in the middle of a legal fight between the Research Corporation and Bell Labs, each of them holding one of my patents. Bell Labs had designed and set up masers to receive microwave signals bounced off a satellite and to test methods of transatlantic communication. Bell Labs was also using masers and lasers in a number of research projects. The Research Corporation sued on grounds that Bell Labs was using the maser patent without a license or permission. Bell Labs refused to pay anything, challenging the patent. This is a typical sort of conflict early in a patent's life. If the owner of the patent does not vigorously defend it, outsiders feel free to help themselves to the relevant technology.

In a sense, the Research Corporation welcomed the opportunity to file a suit. If it could defend the patent against the Bell System, a truly huge fish, all of the other fish in the corporate sea would probably school up and pay royalties. The Research Corporation hoped Bell Labs would cave in, perhaps out of friendship for me, and that would be that.

But before things got that far, the big shots at Bell Labs decided to try to make the whole issue just disappear. Bill Baker, president of Bell Labs, had entered Bell Labs the same year as I and was an old friend. When we happened to both be in Washington, he invited me to join him for dinner at the Cosmos Club. He was most gracious. He said to me, as I recall, "We in Bell Labs have been thinking that we are so appreciative of your work, we would like to give you a prize of $100,000." That seemed marvelous; but in the subsequent discussion he explained that he regarded the Research Corporation's suit against Bell Labs over use of a patented research technique to be something of a travesty on the necessarily open nature of research. He hoped I would understand and have the Research Corporation withdraw said suit.

He was very friendly. I told him the prize sounded very nice; I was appreciative of and sympathetic to openness in research. But I knew that the Research Corporation wanted this issue put behind it in court. If it could not do so with a suit against Bell Labs, it would just have to file suit somewhere else until the maser patent was armored with solid legal precedent. Bill increased the argument a bit by adding that it hurt Bell Labs' feelings particularly to be sued over using the laser, a device for which it felt it could take substantial credit, and which it was only using for research purposes. Well, I told him, the Research Corporation had the patent, not I, and I did not think I should get in their way. The fact is, I suppose, that if I had agreed to talk with the Research Corporation lawyers and ask them to back off from Bell Labs, they might have done so. Yet I felt I could not properly interfere. I did not get the $100,000 prize.

A bit later I ran into a vice-president of AT&T, whom I knew somewhat, and he said he was going to try to persuade the company to settle out of court rather than waste a lot of money on a defense. He did it. The Research Corporation gave Bell Labs permission to keep using masers and lasers for a fee plus later royalties. The settlement provided the Research Corporation with some income, but it did not give it the definitive court precedent it needed to defend the patent against other infringements. That came in the next suit, which was filed by the Research Corporation against Spectra Physics, a company in Mountain View, California, that had become the leading manufacturer of lasers. By then the Research Corporation's lawyer, Harold Stowell, had retired and had been replaced by Dana Raymond, the patent lawyer who handled the late Major Armstrong's case and won it for his heirs. His taking over the patent case was probably fortunate.

I was a bit prepared for a tough fight, because one of my physicist friends, Mel Schwartz at Stanford, knew the Spectra Physics lawyer, who had told him that his company was sure to win. What happened next illustrates the maddening, fine-grained contrariness of patent law, and the sorts of lawyerly guile one needs to navigate its network of technicalities.

Spectra Physics had, for one thing, learned about some digging that Bell Labs' lawyers had done before they elected to settle with the Research Corporation. A surprising legal assertion emerged: the maser patent was no good because the Research Corporation had not filed for it within a year of my first publication of the concept. This was news to me. The first paper I had published with my associates at Columbia was in 1954, and the patent application had beaten the deadline date one year later.

Bell Labs' legal researchers had learned about the quarterly reports we had prepared at the Columbia Radiation Laboratory. The first one describing the maser was dated December 1951. It was an internal document, a progress report to the administrators of the Joint Services research grants. In it were calculations showing that a maser was possible, as well as general descriptions of what the device would look like. It was circulated to other laboratories getting such grants to stimulate the cross-fertilization of ideas.

The quarterly reports were not officially a publication. How could they be? They were not public but internal reports not meant for general circulation, so no problem. In fact, we had sent our report to anybody who asked for it. There had been no effort at secrecy. This publication ruse was all a technicality, of course, a splitting of semantic hairs to suit the law. Still, it was not a publication by the usual definition. That seemed clear to us.

The Research Corporation's stance that this was not a publication may have been unassailable except for one thing. Bell Labs lawyers had learned that a copy of our report had been placed on the shelf of the Harvard University library. Anybody could walk in and read it. This was not our doing, nor had I even known about it; but Spectra Physics asserted that it constituted open publication. If they were right, the Research Corporation patent would fall apart.

To counter a technicality, the Research Corporation needed a technicality of its own. Their lawyer, Dana Raymond, asked me privately if there were any changes at all that we had made between the description in the 1951 Columbia University report and the device that we eventually got to work and which we described in the 1954 paper. The question was whether the Columbia University reports had described a workable maser completely enough. I thought of at least one small change made late in the game. The original design description included a ringlike component at the end of the maser resonant cavity. It was not essential and even gave some trouble as mentioned earlier, so Jim Gordon took it off. Most important, the maser had not worked before we removed it, and did work after that. "Perfect," my lawyer said. This meant that the device described in our early reports was not the one in the patent. Amazingly, this argument and Dana Raymond's skill settled the problem. While the report's presence on the open library shelf constituted legal publication, it had not from a legal point of view provided a complete revelation of the patent. One wonders now, should the later small changes have made Jim Gordon a co-inventor?

Several aspects of this episode show how patent law and common sense can be at odds. For one, the presence or absence of that ring was irrelevant to the essential maser concept, or to later working masers, yet it made a fundamental difference to the lawyers. Second, it seems irrational that a decision by somebody at the Harvard library whether to put a report on the open shelf—a report that was floating around widely—could determine the ownership of a patent. I had no control over what went on the library's shelves, after all. And finally, most interested physicists knew about our maser long before the official publication of 1954; some had read our quarterly reports, some visited my lab to see the maser, and some had heard me give talks about it—but none of these were legal publications and hence had no legal bearing on "public knowledge" about the maser.

Spectra Physics did not fold their tent after losing that round. Their next tactic was to bring up the work of Joseph Weber from the University of Maryland. Weber had published a short paper on microwave amplification by molecules in late 1953 in an electrical engineering journal. It was primarily a paper exercise. It had no feedback or cavity, but it did propose amplification by stimulated radiation. For a time in the 1950s, Joe was rather angry with me. He felt he had first published the maser idea and was not receiving appropriate credit. To ease the problem, we published a joint account of maser history. He had indeed made an early and original proposal; Joe had envisaged amplification by stimulated emission and realized it would be coherent, a property apparently not recognized by a number of still earlier mentions of this phenomenon. Yet his description had some serious practical limitations which I did not point out until asked by the opposing patent lawyer. In fact the amplification that Weber's system might have given was minuscule, and he had not allowed for losses in input or output windows, which would have completely killed this tiny amplification. In addition to that, the application of electric fields he proposed on ammonia molecules would not have produced the particular Stark effects, or splitting of spectral lines, he assumed, and the strong electric fields were so strong that they would have ionized all the gas, destroying the molecules. When the patent lawyers asked why Joe's proposal should not represent a first publication of the maser invention, a couple of years before our patent application, I noted those problems. Spectra Physics gave up on that tack, and Joe himself was quite friendly, perhaps in part because I previously had recognized his very real originality but had not brought up these problems.

The Research Corporation won its case against Spectra Physics, which was settled after the above arguments by consent decree. Thus the basic maser patent lived out the rest of its 17 years, to 1976, relatively unchallenged. Its modest revenue stream at times paid me two or three times my salary, but most of it went to the Research Corporation. Part went into a special fund for grants for researchers around the country, to be selected by a board on which I served. Seeing the money put to good use was indeed a pleasant business. Among the grants I enjoyed was one for $32,800 to Lon B. Knight, a chemistry professor at my old school, Furman University. Another went to my former student Pat Thaddeus, who became a professor at Harvard, to help build an antenna that was to make the first extensive survey of carbon monoxide (CO) gas in our galaxy. Another grant went to the American Association of Variable Star Observers, a group of astronomers centered in Massachusetts, for them to buy their first computer. Recently I had a chance to feel grateful to them in return when I used some of their data in my own work in astronomy.

Bell Labs' laser patent was largely successful, but could have been better. It stood the court test of its central provisions but did not cover everything it might have. The first challenge to the laser patent came from Gordon Gould, the former Columbia University graduate student who had come to me in 1957 for advice about patents, and to whom I had mentioned my conclusion that optical masers—or lasers—could be made.

My impression of Gordon Gould is that every time he learned I was doing something, he would dash to try and cover as much as he could think of in his own notebook. His first record was on November 13, 1957, a few weeks after our first conversation about optical masers; this is the notebook that he had notarized at a local candy store. It included a description of a laser and a list of possible uses for such a device. The next time he produced any substantial written material on the laser was shortly after he obtained a copy of the manuscript by Schawlow and myself, about a year later—but each time, his material was substantial.

Sometime after Gould went to work in 1958 at TRG, the company financed a challenge to the Bell Labs laser patent. It produced Gould's 1957 notebook as supposed evidence that he was ahead of us. There were in fact a number of interesting things in that 1957 notebook, including a fairly good description of a laser with parallel plates, set up like a Fabry–Perot interferometer. But the court examined his notebook and our laser patent, and in the end rejected his claim on two substantial counts.

The first count was that Gould did not really have the pertinent invention in his notebook. It did not indicate whether a tube surrounding the laser was transparent—as we had it—to suppress the multitude of off-axis modes. It also did not say anything about mode control, or just how the laser could be made to produce a single or directional mode of oscillation and, hence, a single wavelength. These were what Schawlow and I had done together, and most of the remaining ideas were in my notebook prior to Gould's conversation with me.

The second count against Gould was that he had not shown "diligence." This is a special patent law requirement, primarily aimed at being sure the supposed inventor really takes his own idea seriously, by either working toward getting it into a practical device or filing a patent application. In essence, the court found that he had not done anything with the idea for quite a while after writing it down and before our publication. Hence he had apparently not been convinced or interested enough in it to make a valid claim. Clearly, the Bell Labs lawyers representing our patent were expert. I'm not so sure Gould's lawyers did as well as they might have; in any case they lost completely.

Gould did not publish any scientific paper in the field until very late in the game, and then only relatively minor ones. It appears to me that his period of quiet and seemingly low interest, during which he could not demonstrate "diligence," lasted just during the time that I was keeping my mouth shut out of a feeling that the idea was Bell Labs property. As already noted, no one other than Art Schawlow and I seems to have been seriously considering laser possibilities during that time. While Art and I were working on our "optical maser" paper, as already indicated, I was careful not to mention it even to my students and close associates at Columbia, such as Ali Javan. Only when Ali took a job at Bell Labs, in the early summer of 1958, did he learn of the idea. Soon after that, he made an important contribution that produced the helium–neon (He–Ne) laser scheme still in very common use.

After TRG's unsuccessful challenge to our basic laser patent using Gould's 1957 notebook, TRG gave up. However, Gould later succeeded in finding other companies willing to gamble on funding the patent pursuit. He was remarkably persistent; clearly, patent rights were something he treasured. He next worked with the Refac Company, an established organization set up specifically to work with inventors and their patents. With Refac, Gould made an effort to claim the patent on the He–Ne laser belonging to Ali Javan, Bill Bennett, and Donald Herriott of Bell Labs, but he again failed. He also challenged a patent owned by the Hughes Research Lab, where the first working laser was put together. This patent was on pulsing of lasers by "*Q*-switching," a method for turning on and off laser action to obtain giant pulses of energy, and again Gould's claims failed. A patent lawyer not associated with the case told me that he thought this was, despite the court's decision, perhaps one of Gould's best-deserved claims but that probably the Hughes lawyers were too skillful.

In the meantime, TRG had obtained a $1 million grant from the new Advanced Research Projects Agency of the Pentagon to try to develop lasers. Theirs was an imaginative proposal, suggesting possible performance and use of lasers considerably beyond what could be clearly expected at the time (though considerably below eventual real laser performance). Still, TRG, in spite of this good funding and hard work, did not produce much early laser success. As already mentioned, one contribution its staff did eventually make was to build an original working model of the helium-pumped cesium laser, which had been suggested in our original paper. I was glad to see a working model of this type built; however, it was never of commercial use.

During the time Gould was at TRG, stories in various popular publications claimed that he was handicapped by being unable to work on lasers himself. It was said that he lacked a security clearance—the result of his having been involved in a Marxist study group while a young student in the late 1940s—and in particular that when the Department of Defense put a security blanket on TRG's laser work, his own notebook was classified. There seems to have been some real confusion about this, even in Gould's thinking. At one point, I obtained a copy of the notebook, and found it was not classified at all. I telephoned Gordon and commented that I had read the newspaper stories that his notebook had been classified. I went on to point out that I had a copy in front of me and that there was no classification stamped on it at all—yet for a document to be classified, such stamps are required. "Well," he said, "you're the first person who has pointed that out to me."

I think Gordon Gould was quite genuinely convinced of the virtue of his case, even when it was somewhat distorted as it was with the claim of a clearance problem, and this may be part of why he was eventually convincing to a jury.

Although my paper with Art had been officially published in December 1958, we had finished it and begun circulating drafts of it to anybody interested by August—as soon as the Bell Labs lawyers had submitted what they thought was an adequate patent application.

Not very long after Art's and my optical maser paper was available, Gould had extensive entries in a second notebook that were dated in November 1958. These entries were to prove the basis for his later, successful patenting of several aspects of the laser. It seemed to me that some of them were indeed already in our paper, but they had not been adequately covered by the Bell Labs patent write-up. This is one of the reasons I regret not having spent more time with the Bell Labs lawyers, to be sure the patent firmly covered the range of technical ideas in the paper.

The company that finally had success on some patents for Gould was Patlex, a then rather new and well-financed outfit formed to work with inventors on patents. It took over the case from Refac. Patlex was headed by Richard Samuel, a smart lawyer, and it had plenty of money. Its tactics seemed to be to push for a number of separate patent claims, to spend considerable effort on publicity, and to avoid suing any of the largest and best-lawyered companies. For such companies, such as General Motors and AT&T, Patlex filed infringement claims, then worked out a fixed-price, out-of-court settlement that was less expensive for those companies than a complex legal defense. In return, the companies got the right to use any patent claim Gould might eventually win in the courts without payment of any further fees, which was clearly a convenient guarantee of no further expenses. Patlex's first trial of a case was against General Photonics, a small and essentially bankrupt firm in California. Patlex won, but I don't think other companies took that very seriously. Patlex then decided on a suit against a small laser firm in Gainesville, Florida, the Control Laser Corporation. Control Laser probably didn't have a patent department of its own, and it hired a local Gainesville lawyer, Robert Duckworth.

Duckworth contacted me as a potential defense witness. I was not eager to be embroiled but felt that if needed I should serve as a witness, and so I did spend a couple of days on the Control Laser case in Gainesville in 1987. I also got a good look at the extra wrinkles in patent law that come with a jury's involvement.

On arrival, I was warned not to be upset by the Patlex lawyer. He was a well-known Florida trial attorney, and I was told that one of his characteristic tactics was to provoke opposing witnesses into losing their tempers. I was told that, in fact, Control Laser had first asked Ted Maiman of Hughes, who had made the first laser, to be a witness. The initial cross questioning so infuriated Ted that he refused to testify and went home.

The jury was, of course, composed of local people. A very innovative tactic that Patlex employed was to pay other local people to sit in the courtroom (unidentified as Patlex employees), to listen carefully, as though they were jurors, and report to the Patlex legal team on how they reacted to arguments and testimony. I understand that ploy was rather novel at the time but has since been used in other cases.

Patlex did quite well over all with the jury. Our basic Bell Labs laser patent was still in effect and not at issue. Also, the jury found that the paper by Schawlow and me had been available to Gould before his critical November 1958 notebook entries, so that it took precedence over the notebook material. Nonetheless, Patlex wound up with important patents. It had a wide variety of miscellaneous patents associated with lasers and their use. Perhaps the three most important items that Patlex won at this critical court case were the following:

1. Light amplification by stimulated emission from atoms or molecules excited by radiation and without reflecting mirrors (supposedly different from our patent because the reflection back and forth, which is normal in most lasers and builds up their amplification, was omitted).
2. Light amplification by stimulated emission from atoms or molecules excited by collisions, again without reflection.
3. The use of laser materials with broad absorption and emission resonances, rather than narrow ones.

To me as a scientist these all seemed unpatentable, since they would be obvious to anyone skilled in the field once the other ideas about lasers were recognized. The first item is essentially the same as the Schawlow–Townes laser patent owned by Bell Labs, and recognized by the jury as being the more senior laser patent but without a resonator to provide laser oscillation. Everyone so far had patented lasers with reflectors that could hence oscillate, or amplify enormously. No one had bothered to separately patent amplification, per se, which went on inside the lasers. The second item was a similar obvious play on the work of Javan, Bennett, and Herriott of Bell Labs, who had patented a laser with mirrors using collisional excitation. Both the first and second items simply left off the reflecting windows, to patent what goes on inside them, the amplification. Gould had previously tried for both of those full patents and lost. Since the reflection of a beam back and forth in a laser gives a large amplification, wouldn't it be obvious to any knowledgeable scientist that there would be some amplification without reflection?

The third item patents amplification with broad, rather than narrow, resonances. There is no clear scientific definition of what is a narrow and what is a broad resonance. Nature provides a continuous variation in width—and no change in principle between any of them. This whole issue would probably have disappeared if the Bell Labs lawyers had included solid-state lasers in the initial patent; they have rather broad resonances, and were discussed in our manuscript, but evidently not thought necessary for the original patent.

The issue of amplification without reflection involved in the last two patents was not new to me. For the early maser patent, recognizing the breadth of the idea, Harold Stowell at the Research Corporation and I debated whether to patent the amplification process, per se. Or, instead, should we just patent the combination of amplification plus resonator, which is what gives more amplification or makes an oscillator. I pointed out to him that beginning as early as the 1920s, several people had mentioned the possibility of amplification in a medium with inverted energy levels and no reflection, and some had devised schemes to make it work although they had never carried them out. In addition, patent law states that anything that occurs naturally cannot be patented. I told Stowell that it was probable that some amplification occasionally occurs spontaneously in explosions or lightning strokes, though no one had yet seen it happen. For those reasons, he felt it would probably be useless to add the amplification process to the patent separately. The idea seemed obvious if oscillators were made. If obvious to persons skilled in the field, it would not be patentable, and probably most applications would use resonators anyhow. Probably the Bell Labs lawyers were thinking along the same lines when they wrote the laser patent, since they didn't bother to patent amplification, per se, although I never discussed this issue with them. Thus, previous lawyers had simply omitted patenting the two most important things that the jury awarded to Patlex. On top of that, Patlex claimed that since lasers with resonators also used amplification, its new patent should also cover normal lasers—which already had been patented!

In an effort to refute the Patlex claim to amplification, during the jury trial I urged an argument that amplification (without mirrors to make a resonance) is a natural process. I suggested to Duckworth that he ask Mike Mumma of NASA's Goddard Laboratory to testify. Mumma's group, as well as ours at Berkeley, had by then discovered that sunlight striking the atmosphere of Mars excited carbon dioxide (CO_2) molecules in such a way that they slightly amplified infrared light. The Goddard group had done a particularly good job proving the scientific case for this. It seemed to me that Mumma was the perfect one to show that amplification is indeed a naturally occurring process and therefore, according to patent law, an unpatentable one.

Mumma carefully told the jury about discovery of natural laser amplification in the Martian atmosphere. The Patlex lawyer's tactic was to make a joke. He rolled his eyes and said, "You mean little green men on Mars?" The jury went along and decided that this example from space didn't count in a court on Earth. I had suggested the tactic, and I wanted it to work, but I don't think the jury's decision to exclude Martian lasers, on a common sense basis, was so unreasonable. Even though still other types of lasers have also been discovered in an astronomical object, natural laser amplification is certainly not present in any obvious way. Yet the other cases I had been in, involving only lawyers and judges rather than citizen jurors, had made me expect strict adherence to patent law, even when it was strange. So I was surprised and amused that this time, a cute technicality which, according to the legal wording should invalidate any patent, didn't work: a jury made the law. On a still more reasonable basis, and forgetting about amplification on Mars, the fact that Patlex obtained a patent on an aspect of lasers that had been apparent all along in the earlier patents—and an essential part of them but simply not previously singled out—did nothing to help the jury decision make much sense to me.

Duckworth, Control Laser's lawyer, had warned me that things would not end simply. He said that whoever lost would want a retrial. He felt the first trial would serve primarily to map out the claims and arguments, while a second trial would be needed to really settle matters. Patlex was either very clever or very lucky. On Patlex's urging, the judge ruled that Control Laser had to pay the claimed patent fees immediately. They were substantial, in the millions of dollars, and Control Laser could not pay them. The company had to turn all its assets over to Patlex. No further challenge was possible. The case was closed.

I believe no company has challenged Patlex's patents in court since that Florida case. The patents have raked in many tens of millions of dollars, as they cover a wide variety of lasers. The total was greatly enhanced by the years of delays to Gould's claims in the patent office and in the courts. My basic patents, issued in 1959, had just expired in 1976, when Gould's patents went into effect. As my lawyer told me early in the game, the most money is made by the patents that take the longest to be issued—and by subsidiary but later patents that may involve only minor modifications. Gould's persistence, coupled with a long series of failures and delays but eventually some success, turned into rewarding good fortune.

Quite a game! The patent process makes little sense in terms of scientific innovation; it is not one in which to be involved if you really enjoy science and not legal wrangling and gambling. Ted Maiman, for instance, deserves much credit for innovation in building the first actual operating laser. Still, Ted has largely been shut out of any widely applicable patents, perhaps in part because the Hughes lawyers did not assume a foresighted, aggressive role.

Today, the laser business is worth many billions of dollars per year. Most of the more important Patlex patents on lasers have also expired as of the mid 1990s. There are other inventors and many other patents on particular types of lasers and their applications that are still in effect, of course, but none has been as contentious or played up in the news media as much as those described here.

And recently things have changed. Patent laws have been modified in two ways, which would have produced a story quite different from what has just been traced out. First, since 1995, a patent now expires 20 years after its application rather than 17 years after it is awarded, in order to shorten long delays. Under this law, the patents that Patlex finally secured would probably have expired within a few years after they were awarded. Second, juries are no longer to be used to determine the scope of patent rights, only judges. In view of the increasing technological complexity of patents, that's probably a good idea.

8

ON MOON DUST, AND OTHER SCIENCE ADVICE

Late in the summer of 1951, I was a member of a party headed for a dry lake bed in the Southern California high desert, west of Death Valley and east of the sharp-edged peaks of the southern Sierra. We were to visit an installation at China Lake, born as a rocket-testing site and run by Caltech during World War II. The Navy ran it after that. We found a cluster of radar tracking sites, hangars, barracks, and laboratories set on an alkaline flat, amid lava flows and cinder cones. The smooth and wide extent of the dry lake bed, like an aircraft carrier without edges, and the isolation from any dense human settlements made it a perfect place for testing rockets, missiles, or experimental airplanes.

The Navy had for some time been supporting infrared research, but up to that point, almost nothing useful to the Navy had come of it. Our party was a scientific advisory committee charged with looking into whether there were really going to be any significant military uses for infrared radiation and related technology. The chairman, Donald F. Hornig, a chemist and infrared spectroscopist at Brown University, had asked me to join shortly after my initiation of the Navy's microwave committee. I was happy to sign on to what was called Project Metcalf, after the Metcalf Research Laboratory at Brown which Don headed. It was a way to get educated about infrared, and it was a pleasure to join the other members of the committee, which included physicists J. H. ("Van") Van Vleck of Harvard, "Pief" Panofsky, Leonard Schiff, and Bob Hofstadter of Stanford, Dale Corson of Cornell, David Dennison of Michigan, chemist Gordon Sutherland of Cambridge University, and engineer-physicist Gene Fubini.

It would also satisfy a favorite pastime of mine: to visit laboratories and see what other scientists were doing.

My career has been focused primarily on scientific research. But this trip to China Lake is an example of another sort of activity that has taken much of my time: providing science advice to the government and, to some extent, to industry.

Science and technology play such an important role in our society, and public decisions about them often require such specialized knowledge, that scientists are more and more required to participate in society's and government's decisions. My own involvement frequently depended on my particular scientific background, as it does for many other scientists, so the maser-laser development frequently showed up and affected my involvement. Intense research and service to government or industry do not mix well, yet any prominent scientist is likely to be called on for such service. I would like to describe a few examples in some detail.

Certainly China Lake, with its stark landscape and clear desert air, was one of the most visually unusual research sites I'd seen. The first day there, we met with Bill McLean, a former classmate of mine at Caltech. Bill was a modest, straightforward person who, as an experimental physicist, was both imaginative and practical. He had been developing an idea of his. It was called Sidewinder, after a small rattlesnake native to the area. A rattlesnake gauges its strike by using the warmth of its prey as a guide, and so did McLean's Sidewinder, which became the prototype of a very important class of air-to-air missiles. It was to have the fearsome ability to home in on the infrared radiation—the heat—from a target's engine and exhaust. Its "eye" was an array of four infrared detectors which, by responding to heat from another plane's engine, could guide the missile left, right, up, or down, to keep it right on track to an interception. The idea had been tested on the ground and seemed to promise great and deadly efficiency.

The Sidewinder was in an early and exciting stage of development when we visited China Lake, and it was what we called an "under the bench" project. That is, McLean had been working on the Sidewinder on his own initiative, with general funds of the station. The preliminary results and the potential of the Sidewinder immediately impressed the committee. Still, it had not been authorized by higher-ups at the Pentagon. And just before we got there, top Navy brass had gotten wind of the Sidewinder. The response was unequivocal: Stop. An air-to-air guided missile, they told the director of research, was not part of the station's assignment, since it was primarily a rocket-development site.

Our committee, drawn from far outside the chain of command, had no reason to care whether the missile fit into the mission statement for China Lake. It seemed clearly a mistake for the Navy to pull the plug. Hornig asked for the names of the admirals responsible, telephoned them back at the Pentagon, and went to Washington to see them. Speaking for the committee, he told them that the Sidewinder program was one of the best things they had going. Some of those top admirals objected that planes had to attack from the front, not from the rear as these heat-seeking missiles would do, aiming at the hot tail pipe of the engine. But Wally Schirra, recently back from successful air fights in Korea and then stationed at China Lake, stood up and said, "Sir, the only planes I've ever seen shot down successfully were shot from the rear." Don was amazed at the young officer's brazen disagreement with the admirals—and has always thought that this event or attitude might have had something to do with Schirra's transferring to NASA and becoming an astronaut.

Our written report later played up the potential of the Sidewinder as the Navy's most promising use of infrared. Faced with such forthright advice, the Navy said well, okay, if we thought it was so important, it could continue. In fact, one of the admirals to whom Don Hornig talked in Washington was "Deke" Parsons, who the same evening flew out to China Lake to see for himself. Within a few days after our visit, McLean was all set to proceed with the Sidewinder.

Today, descendants of those early Sidewinder missiles remain among the most effective weapons carried by Navy, as well as Air Force, combat aircraft. The near cancellation of the missile came more from the Navy's respect for organization charts and budget plans than from basic technical judgment. The Pentagon admirals were simply following their system, but they responded sensibly when given an informed outside point of view. The final Project Metcalf report covered possible applications of infrared quite broadly; it involved a great deal more than the Sidewinder, but the reactivation of this missile development was a highlight.

If only it were always so easy to give scientific advice—and to have it followed! The science adviser's job is loaded with political undercurrents and traps. Yet, almost any scientist who gains some prominence is likely, sooner or later, to be asked for an opinion by a government agency. We regularly get asked for evaluation and advice, frequently on concerns that extend deeply into government policy. The pattern of advice and response is rarely as straightforward and positive as was the case for the Sidewinder.

Individuals employed in political science, sociology, economics, or other fields pertinent to government decisions—particularly university faculty members—often ask why scientists are so specially called on or "privileged" as consultants to government. There is a very human reason. Politicians don't think they need much advice from political scientists and other such people. After all, as professional governors, politicians usually regard themselves as the ones who really know their business of social and political issues.

Science is something that most political figures know little about. Yet, and especially since World War II, Congress and the executive branch have found themselves forced to allocate a great deal of money for science and technology—often directly for the military or for expected economic development. In part, they admire science for its intellectual value. More important, they believe that science helps produce both economic and military success for the United States. What other intellectual or academic areas could possibly obtain Washington's support on the scale given to science and technology? There are endowments and so forth for the arts and humanities because they are culturally interesting—even uplifting—but there is no direct payoff that Congress sees. So, because politicians must make important decisions about science and technology and must direct so much money to them—yet often recognize that they understand them so little—they can be hungry for advice. Their attitude is often shared by high-ranking military officers. As an ironic result, social scientists, the academic people who are the most professional in mastering the nuances of public policy and who may seem most inclined and prepared to give advice to government, are perhaps called on much less than are scientists by the White House, Capitol Hill, and the Pentagon.

Of course, one cannot escape politics when working for the government. Unless one is really way out on one side or the other, personal politics don't interfere a great deal with advising on science or technology. I consider myself a Southern liberal: financially conservative but rather liberal on social policy. That makes me a middle-of-the-roader. I don't feel fierce or dogmatic allegiance to any one group, and play it as I see it on issues. I have been asked for advice by as many Republicans as Democrats.

Through the 1950s, I spent a modest amount of time on such advisory groups, particularly including work for the military, NASA, and the White House. If a committee was concerned with matters that were more or less in my field, I was also interested and felt I would probably learn something.

In spite of substantial work on governmental issues, I have never been emotionally caught up in government policy. My attention to government problems has stemmed more from a sense of public duty. This may have been an asset. I could look at most of them without prejudgment and simply try to figure out the best thing to do. The distance between governmental problems and my personal career or emotions has also meant that while I would work intensively on a government task, after it was over I often forgot most of the details. Scientific work and events were much more in my mind and are easier for me to remember in detail. In any case, I always got back to scientific research happily if my advice or management efforts were not wanted or did not seem to be particularly important.

After several experiences on part-time committees, my in-depth plunge into government science policy and a close look at how government works came in 1959, just about the time of the quantum electronics meeting in New York that preceded the first successful laser.

It was a particularly tense phase of the Cold War. Fresh in everybody's mind was *Sputnik*. When that 84-pound Soviet satellite rose into orbit in October 1957, and beeped its way over American skies, it set off loud alarms from both the public and government, especially the military. Nobody was sure what was going on behind the Iron Curtain, but the satellite looked like unwelcome news. It raised the fear that the Soviets might be dangerously ahead on intercontinental missiles. By the end of 1959, the Soviets had even sent a rocket to the moon. It seemed the Soviets were very advanced in space rocketry and, more important to our national defense, could build missiles with nuclear warheads that might hit the United States from the middle of Russia. The Soviet love for secrecy only made the shadowy glimpses of technological developments in the Soviet Union seem more ominous. For the first time since the end of World War II, the United States felt uncertain of its ability to adequately control international military pressure. Also for the first time since the war, our technology was not fully dominant. The Joe (Senator Joseph) McCarthy hearings, with their distorted efforts to root communism from American institutions, had just subsided earlier in 1957, when along came this new fright from the ideologically alien adversary whose leaders had sworn to bury us technologically and economically.

The McCarthy excesses were bad enough, with the institution of loyalty oaths, and my students' consequent fear of even studying Russian, because they might be accused of being communist and subject to severe political attacks through the application of "pinko" and similar labels. Such events were mostly our own fault and of course we would come to our senses in time. But this external threat, of intercontinental ballistic missiles and space weapons, might be real.

I would not say that there was wide panic. President Eisenhower, for one, tried to reassure the public. In his speeches he declared that we were not really behind in missiles or space research. Yet he was not very successful at convincing the public—or his political opponents. He knew he had to give such matters urgent, expert, and prominent attention. To help him follow and appreciate the technological and scientific situation, Eisenhower set up the President's Science Advisory Committee, or PSAC. It followed the tradition of science advice that had been established by Vannevar Bush in World War II, but was on a more formal and official basis. Its head, and the first of the modern presidential science advisers, was Jim Killian, the president of MIT. Killian was not a technical man, but he had a good sense of how to put together scientific talent and to arrive at balanced advice.

In early 1959, I was at Columbia, fascinated with the science of molecules, atoms, and nuclei. I was also preparing to build a working laser. Like a bolt from the blue I got a call from Garrison Norton, a former prominent business accountant. He was well acquainted with people in Washington and had become head of a nonprofit agency—one organized by a group of university presidents to assist the government—called the Institute for Defense Analysis, or IDA. Norton came up to New York and asked me to be IDA's vice-president and director of research. I'm not sure why he asked me, except that I had done some work on military advisory committees and, by then, was becoming fairly well known in scientific circles. I am sure it wasn't any specific scientific work I had done, because Norton did not follow highly technical issues closely and certainly knew little about the maser or the exciting prospect of an optical maser, or laser.

It was an honor to be asked. Nonetheless, I knew they were kind of desperate. John Wheeler, a distinguished physicist at Princeton University, had already turned the job down. Wheeler and I knew each other pretty well. He may perhaps have recommended me for the IDA job.

Some of my colleagues warned me in various ways, "Don't go down there. Stay and build the laser. That is the work that will get you the Nobel Prize." I thought the maser and laser might, in fact, win a Nobel. But, I felt it did not really matter who actually built the first one. The ideas were there. I was not going to make a career decision to go all-out to build one just to win the prize.

Nevertheless, it was a busy and interesting time for me in the lab at Columbia, and the Washington job did not sound like any fun. I. I. Rabi told me that I must think Washington was in desperate condition to even consider going there. I talked things over with Frances and the children and told them I felt a sense of obligation. While a good number of senior scientists, such as Rabi, were called on regularly for advice, it seemed to me that scientists of my generation needed to start pitching in. Perhaps my usual interest in trying new things also played a role.

There were only a few top-ranking scientists then working full time in Washington, which was part of the reason I felt strongly that more scientists were needed. Herb York, an exception, had surprised a lot of his colleagues by leaving the Lawrence Radiation Laboratory in Berkeley to be chief scientist at the new Advanced Research Projects Agency (ARPA), a small but independent branch of the military set up in response to the recent Soviet successes. Its role was explicitly to explore new, often secret weapons technologies. I remember asking Herb at about that time why he did it, and he joked, "Well, they paid me a whale of a lot of money!" It was a way of saying that whatever the reasons for working in Washington, enjoying science or enhancing a professional reputation were not among them.

The trustees of IDA were largely university presidents and similarly prominent, public-spirited people. Circumstances, particularly the *Sputnik* scare, had made Washington hungry for technical advice and in the mood to call on and trust scientists. IDA seemed in position to be very, very influential. I felt I could stand it for two years. And so, I went to Washington in the fall of 1959.

We found a modest house in Washington that suited us well, in a neighborhood that seemed right for the children. I was particularly pleased that it was a house in which President Woodrow Wilson had lived briefly. Washington did indeed have its excitement and interest. There were many social and other contacts with government personnel. Cocktail parties seemed to be remarkably important for friendly relations, informal deals, or scuttlebutt—and I became a bit fed up with that—but it was an interesting city. In the middle of our stay, there was the excitement of the start of the Kennedy era, with a number of new academic types entering government. We had the opportunity to attend the inaugural ball and, of course, we went.

My office was in a nice building about three blocks from the White House, with expert secretarial and administrative help. My primary job was to get good personnel to advise on the new technical developments bursting on the scene, and to monitor the findings and quality of IDA's research and analysis studies.

Shortly after I arrived on the job, Director of Central Intelligence Allen Dulles invited me to a briefing. It was, as befits the Central Intelligence Agency (CIA), in a nondescript building in a nondescript part of town, with heavy security and a half-dozen top-level intelligence types in attendance. The topic was the CIA's information on Soviet intercontinental ballistic missiles (ICBMs). Of course, this was not something about which I had any particular authority, but I was glad to be informed. They seemed to tell me everything they knew.

The briefing went on most of the morning. As it ended, Dulles turned to me and asked, "What do you think? How many ICBMs do you believe the Russians have?" This struck me as a surprising question to be asked. All the special information I knew was what his briefing had just told me. And I was wondering about the same thing. What did the total of our information have to say?

I had to tell Dulles I believed we did not know. It could have been anywhere from zero to 100. I recall saying that the only thing we could do was to establish an upper limit to how many they might have produced—and, while they may have built a good many, they may or may not work well. The affair, in which the CIA director asked me from a cold start to do what his analysts presumably could not do after many months of studying the question, simply indicates how worried he was and, perhaps, how inflated their opinions were about the power of scientists to give fast and wise answers. Today, thanks to the space program and satellite surveillance, we can answer such questions well and fairly surely. Even then, the truth became clearer rather quickly. President John F. Kennedy learned soon after his election that the "missile gap," which he had used in political speeches to discredit the Eisenhower–Nixon administration, was exaggerated. Our U2-planes, high-flying reconnaisance aircraft, showed that at that time the Soviets possessed no intercontinental ballistic missiles that could be a real threat to the United States.

About eight months after I arrived in Washington, Ted Maiman at Hughes had the first laser working, and other types of lasers followed soon after. Unavoidably, I was called upon to advise on military-sponsored laser research, and to answer the considerable doubts (as well as overblown hopes) about their military usefulness. One of the laser-oriented scientists I recruited to spend some time at IDA, Robert Collins, was particularly downbeat on lasers for the military. Of course, I could not be certain what their potential was either, but I was sure they were worth investigating. Collins asked, at one point, whether I realized that to pump a laser up enough to fire it at a missile with any effect, "You would have to have the energy in a pile of dynamite as big as a skyscraper." Similarly, at a big meeting on lasers, Eugene Fubini, an engineer and then an assistant secretary of defense, said lasers were far too inefficient to ever produce enough power to interest the military. He was right for lasers of that time. I did not have any great faith in the laser as a highly destructive weapon myself, but I had to make the point that there was nothing inherently inefficient in laser operation and that, in principle, a laser could convert energy as efficiently as a steam engine or any other practical device. It was just a question of finding the right techniques. So, the work continued. While there are still no good killer-ray-gun lasers in operation for the military, and there may never be, highly efficient lasers are in fact now common and some large ones can melt metal at a distance of miles.

It was hard work, sitting through long meetings and overseeing the various committees and groups within IDA. I got little science done myself, but there were a few exceptions. For instance, Philip Morrison and G. Cocconi of Cornell University had just written a paper on listening for radio signals from extraterrestrial civilizations, which interested me. They felt a logical frequency would be in the microwave region of the spectrum, near the 21-centimeter line of neutral hydrogen. I calculated that it could just as easily be done with lasers in the visible or the infrared part of the spectrum, and wrote a paper with R. N. Schwartz, a young scientist at IDA, on possible detection of alien laser signals. The paper arose partly out of the kinds of technical questions facing IDA, but it was also science.

Most of the time, I found myself working in thickets of acronyms representing interlocked agencies and committees. PSAC, the President's Science Advisory Committee that George Kistiakowsky had by then taken over from Killian, was the top science advisory body, and George welcomed my occasional participation there. ARPA, the Advanced Research Projects Agency, was pushing high technology and space work forward and answered directly to the secretary of defense (its head at the time was formerly head of General Electric's refrigerator division, not a technical person). IDA, a nonprofit institution set up for public purposes, hired and employed the scientists who were actually located in the Pentagon, made the technical decisions for ARPA, and passed them on to the relatively few government officials in ARPA.

Technological successes were highly prized in this network of agencies. Shortly after I had arrived in Washington, the United States sent up a satellite which flew over Asia and for a few minutes broadcast a message from President Dwight D. Eisenhower to the world. It had been planned by ARPA personnel as a demonstration of American technical parity with *Sputnik*. They were all delighted that it worked. How much of the world was tuned in to the rocket's broadcast during that short time and heard Eisenhower's voice, or how much good it did, is questionable, but the episode illustrated the tense and worried spirit of the time.

Another important group that IDA maintained for the military was the Weapons Systems Evaluation Group, or WSEG, whose members were civilian scientists placed in the heart of the Pentagon to advise on which new weapons and tactics would probably work and how well. There was a lot with which the military wanted help, including space technology, submarine detection, new materials, and new radar-related schemes. The generals and admirals also wanted outside experts to help them evaluate proposals—which often amounted to sales pitches—from industry for new weapons systems that used esoteric new technologies.

The cooperative spirit of the time was somewhat unusual. When new technical problems arise, and when new government personnel arrive on the job, that is the time when good technical advice can be most effective. As time goes by, the leaders of any enterprise become surer of themselves and less and less willing to listen to outsiders. Policies and opinions get frozen in place. Such cycles of trust and distrust are perfectly understandable, and I've been through several of them.

A striking example of openness when there is a fresh start came when I was asked to chair a committee on plans for the space effort, as one of a number of committees reporting to the newly elected President Richard M. Nixon, on policies for his term of office. Nixon met with all the committee chairmen and noted that it turned out that more Democrats were among the committee membership than Republicans, saying that this was the kind of openness and interaction he wanted during his service as president. That view seemed quite genuine at the time, but of course was to change during the course of President Nixon's term.

An example of such a shift toward a desire to control, and away from openness, came toward the latter part of my stay in Washington. I was sitting in a meeting of IDA advisers to the Weapons Systems Evaluation Group, when the general at the head of the table told me I should leave. While the group's members were screened, hired, and paid by IDA—by me, in effect—he was the official head and said I did not belong. There I was, vice-president of the IDA, which was supposed to supervise these scientists, but with the door shut in my face. Well, I had not been all that eager to stay in Washington in the first place and was not of a mind to accept the general's demand. I made a pointed objection to Pentagon officials. I said, "If we are supposed to be supervising these people, how can we do it if I can never listen to the discussions? How can I be effective?"

Initially, I got my way, but some of the military kept up the pressure until we decided that IDA should just bow out of the Weapons Systems Evaluation Group. The generals wanted scientific advice, all right, but they wanted to hire the scientists directly and put them under their command, so that the generals' own ideas could be more dominant. That was a natural, human reaction and is gradually what happened to much of IDA's role in providing technical advice. More and more scientists were hired directly by the Pentagon, until the brass felt they no longer needed outside expertise. The IDA is still effective and useful, but has gradually been reduced to one of many Washington think tanks.

The IDA's most important, lasting contribution during those years may well have been an institution called Jason, founded in 1960. Today, the primary regular Jason function is an extended gathering of academic scientists, plus a few representatives from industry, each summer in San Diego, with a few yearly meetings in Washington. Topics at these gatherings, some of which are highly classified in nature, include military systems, intelligence-gathering tools, arms control, nuclear energy, atmospheric pollution and greenhouse effects, the space program, controlling chemical and biological warfare, and the technological aspects of diplomatic issues. Jason has a unique and very influential standing in the world of government science advice. Its core is a group of exceptionally capable academic scientists who are interested in contributing to public affairs; they work hard toward the objective evaluation of national issues involving science and technology. Jason was begun with a group of outstanding young scientists, but now, with the passage of time and the continued appointment of new young members, its members include a wide age range.

We did not invent Jason from scratch. For a few years before I had arrived at IDA, a set of bright young scientists had been gathering each summer at the Los Alamos National Laboratory and consulting on military problems. Somewhat separately, but with an approach rather similar to what Jason was to adopt, John Wheeler, Eugene Wigner, and Oscar Morgenstern of Princeton had organized "Project 137," in 1958, involving a group of academic scientists getting educated about governmental problems. The gathering at Los Alamos had the attractive atmosphere of a working vacation for participants, who typically took their families. Los Alamos gave them a salary, and they discussed scientific and technical issues, mostly with government, military, and national security implications. Regulars in the group included Murray Gell-Mann of Caltech, Ken Watson of the University of California at Berkeley, Marvin (Murph) Goldberger of Princeton, and Keith Brueckner of the University of California at San Diego.

I learned about the group in early 1960 from Marvin Stern, a young mathematician on the West Coast. He told me that its members were thinking of setting up a company, which would permit them to make further plans for this type of work and to run the affair themselves, rather than depending on Los Alamos, the Atomic Energy Commission, or industrial companies. That looked to me like a perfect thing to be adopted by IDA, in the form of a publicly oriented nonprofit group. The Los Alamos operation had attracted just the sort of first-rate young scientists I felt should be involved in advising government. When the older generation of scientists that government had been calling on since the war was gone, it would be vital to have another group familiar with government problems, ready to come into government circles, yet with the independence provided by academic positions. I took the idea to Garrison Norton. He bought it right away. Then we both went over to see Secretary of Defense Thomas S. Gates, and he agreed with us.

In addition to the Los Alamos gatherings, there were other ad hoc summer study groups set up, from time to time, to look at particular problems, and they provided other precedents for Jason. Probably the first was one organized for the Navy by Jerrold Zacharias of MIT, primarily to study detection of submarines. But before the Jasons, there had been no permanent or continuing groups of academic scientists that could become familiar with a wide variety of government problems and, at the same time, be truly independent outsiders, unbiased by significant dependence on government pay.

A big problem was getting security clearances for members of the group. The military, of course, is full of restrictions on information and tends to keep projects compartmentalized so that people don't know any more secrets than necessary to get their work done. The Jasons, by their nature, would tackle broad, sweeping questions that might touch on many separate programs, each of them in an institutionally isolated enclave. It took a while, but the intense feeling of need in Washington for the best of scientific advice finally allowed us to work out an excellent arrangement that gave potential access to almost anything.

Once we explained to Pentagon leaders the basic philosophy of the effort, most of the individual clearances came automatically—but not all. For instance, it took the military administrators a while to accept Murray Gell-Mann. He was (and is) an outspoken man, who did not mind pointing out ideas he regarded as foolish. Like many academic scientists, he was not always highly respectful of government. A few people in the Pentagon found it hard to accept that it is possible to be a good, reliable citizen and also be so highly critical of accepted policies. They often saw things in rather stark terms. To them, some of the Jasons were academic liberals who were a bit too irresponsible and tolerant of left-wing ideas.

We set up the Jasons with a governing structure that included a steering committee, with Goldberger the first chairman. To give the group further credibility and experienced advice, we included four senior advisers: John Wheeler and Eugene Wigner from Princeton, Edward Teller from the University of California, and Hans Bethe from Cornell. Teller, not then in quite as much disrepute among liberals as now, represented the conservative side, while Bethe was more liberal. I purposely tried to get a spectrum of points of view. The senior advisers were to meet regularly with the group, and they and the other 20 or so members would work at government problems without having to leave academic life. Since then, the senior advisers have been dropped, but the basic idea has worked well. The members come to meetings on various topics, hear briefings, discuss possibilities among themselves, work on the problems at home some more if they like, and have extended, focused studies each summer. They also get a good consulting fee.

We got started by the summer of 1960, with the first meeting that summer at Berkeley. Later, the summer meetings were and continue to be regularly held in San Diego. Initially, I represented the IDA administration, but later was simply a member within the Jason group. IDA helped set up and adjust the needed administrative arrangements. A few years later, the nonprofit corporation called MITRE took over the administrative functions of Jason.

Jason has remained remarkably independent and effective. As hoped from the beginning, Jason members have gone on to serve in other roles; a number have been members of PSAC, the President's Science Advisory Committee, and of advisory bodies to Congress, NASA, the CIA, or for various arms control activities. As a measure of the independence of the Jason group, one of its members, Dick Garwin, has been very busy in recent years as an aggressive critic of government policy and has been harsh on the build up of particular U.S. weapons that he considered either destabilizing or wasteful. His frequently sharp-edged public positions have not always made him a favored adviser in military circles, but he is a Jason member, very knowledgeable, and active on a wide variety of issues. His important contributions were recently recognized by the R. V. Jones Intelligence Award of the CIA.

At one point, the head of ARPA (or DARPA, the Defense Advanced Research Projects Agency, as it was called later) tried to push the Jasons more under direct government control. He wanted them to work only on problems assigned by DARPA or he would not give them money. I went with several others who have worked with the Jasons for a long time and talked to Undersecretary of Defense Don Atwood, whom I had known and admired when he was at General Motors. He seems to have blocked that misdirected effort.

The Jasons have contributed importantly in ideas and analysis and, today, have such a good reputation, visibility, and so many friends that the organization seems reasonably well protected from government officials who might prefer to restrict its free-ranging operation and objectivity.

Inevitably, the war in Vietnam, which posed both ethical and technical problems, engaged the Jasons. One Jason group's proposal came to be called the "McNamara Wall." It was envisioned as a system of electronic sensors and other devices that would seal the border of South Vietnam against both infiltration and the resupply of Communist forces from North Vietnam. It was very appealing to Defense Secretary Robert McNamara, who loved advanced technology. He endorsed the Jasons' plan enthusiastically and as a result it became associated in the public's mind with him.

The Jasons, as a whole, were not very sympathetic with U.S. policy in Vietnam. They did like the idea of perhaps reducing the use of bombs and missiles, and the scale of the fighting, by using modern electronic devices to help monitor the border. George Kistiakowsky, although he was not then formally a Jason, got involved and eventually became chairman of the Department of Defense committee on the McNamara Wall, a committee composed largely of Jasons. Kistiakowsky backed the Wall as a good alternative to pure brute force. Of course, the Wall remained an instrument of war. After its sensors detected significant movement by the Communist forces, an attack by artillery, aircraft, or ground forces was to follow to stop any infiltration from the North. The idea was that the electronics would be so good at detection that supplies to the North Vietnamese troops in South Vietnam and also to the Viet Cong would drop well down, reducing loss of life on both sides.

Some of the military high command were not enthusiastic about the McNamara Wall. Perhaps from their vantage point, the Wall was the product of an impractical bunch of eggheads, with McNamara chief egghead. Their skepticism slowed down efforts to try the idea on any large scale.

By that time, I had left IDA and gone to MIT to be provost. While I did not help develop the Wall idea, I reconnected with the Jasons just as debate over issues stemming from the McNamara Wall was reaching a head. It was not just the Wall causing trouble. The Jasons were also getting political flak from friends and colleagues opposed to governmental and military policy. Many Jasons worked at universities whose students and faculty members were largely hostile to anything seen as supporting the war in Vietnam. A few of my Jason friends, including Henry Foley from Columbia, felt threatened on campus and were worried about their families and their offices. They had been sticking with it for quite a while, but about the time I was reconnecting, a concern was growing that perhaps it was time to disband.

The problem was not just that the Jasons couldn't take the heat. Several members honestly felt ethically uncomfortable providing advice that would shore up the government's efforts in Vietnam. I understood. The opinion that the war had become a sad mistake was certainly one that I shared, but I felt that we needed to work out a solution to the very damaging course of events both for the United States and for Vietnam.

One day, we had a full discussion around a table with most of the Jasons debating whether to call it quits. Because the feelings were tense, I remember my own plea. "You know we've got lots of Americans over there, along with Vietnamese, all struggling. Their lives are at risk, and it seems to me that we ought to try to help regardless of what our friends think. They've got their lives at stake and all I have at stake is my time, my friends, and my reputation. I'm willing to risk that."

A few did resign, but most hung in there. Among those who pulled out of government advising was Kistiakowski. Overall, I think it was a good thing to have a few leave, while most kept working to give the best advice we could. George made his resignation quite public, saying he was never again going to have anything to do with the military. I had a great deal of respect for him. That sort of public opposition probably helped change

American opinion about the war. At the same time, it would have been a bad thing to have all of us walk out.

With Kistiakowski gone, I was asked to chair the McNamara Wall committee. One of the first persons I consulted for advice was Kistiakowski. He thought I should do it, saying that the committee was needed even though he had felt compelled to make a public protest. Paul Nitze was Undersecretary of Defense at the time, and I also went to him for advice. I told him, "Look, I'm not in favor of what the country's doing. We need a way to limit killing and still settle the war. We certainly ought not escalate by heavy bombing of cities in North Vietnam." Nitze said he basically agreed, but that he could not go on record opposing the president's policy. I told him that as long as his goal was essentially the same as mine, I would serve on the committee.

I asked if I could talk with Defense Secretary Clark Clifford. Nitze told me Clifford was not seeing anybody on the issue of bombing North Vietnam, but he implied that Clifford also was leaning against it. Before long, in fact, the bombing stopped under President Johnson's orders, and I do believe Clifford was instrumental in that.

While it seemed to me best to continue providing inside advice, there were limits. I did not last on the McNamara Wall committee very long myself. I met with the committee about two times. An Air Force general, Jack Lavelle, was there as liaison from the Pentagon. It was clear that he was against the McNamara Wall as the Jasons had proposed it, no matter

Figure 12. The President's Science Advisory Committee (PSAC) meets with President Lyndon B. Johnson in the cabinet room of the White House, 1966. Johnson is at top center and had just made a joke. To his left are Herbert York, Charles Townes, and Lewis Branscomb. To his right are George Pake, Philip Handler, and Sidney Drell.

what McNamara wanted. Lavelle wanted to use the Jason-conceived technologies, but not to help seal the border. There seemed no hope of putting the project through, so I resigned. The committee sort of fell apart. That particular general clearly had felt strongly and, apparently, also had a hard time following orders. Some months later, he was court-martialled for organizing and leading a bombing raid on North Vietnam, after Johnson had ordered them stopped.

Interestingly, the military did put the McNamara Wall sensors to good use. During the Tet offensive in early 1968, about 6,000 U.S. marines were besieged at Khe Sanh. North Vietnamese units were all around them. The newspapers had built it up as a potential replay of Dien Bien Phu, where French forces were defeated in a long siege that led directly to France's pullout from their Southeast Asian colony of French Indochina in the late 1950s. Khe Sanh had the potential to be a massacre, but the American forces survived. David Griggs, a geologist who became science advisor to the military in Vietnam, told me that at his suggestion the Air Force dropped a large number of the Jason-designed sensors around the garrison. They gave such precise information about North Vietnamese movements that American artillery was able to break up any massing of forces, and the attack was abandoned.

At one point, I was asked to be the overall science adviser on the Vietnam War. It would have meant moving with my family to headquarters in Hawaii, and put me in a position in which I could not clearly oppose the administration's policies in pursuing the war. That was just out of the question. I wanted to be helpful, but I did not want to have to reinforce or be a mouthpiece for the administration's actions. As a lower level and independent adviser, I was free to tell anybody when I opposed anything, and why. If they didn't like it, I could simply disconnect and spend more time on my research. To take a leadership position, where I would sometimes have to defend government policies even if I fundamentally disagreed, was impossible.

Being prominently associated with the Jasons at that time caused a number of uncomfortable incidents. A bit after the affair with the McNamara Wall, but with the Vietnam war still going, I happened to be in Italy attending an international scientific meeting. Activists in Europe associated the Jasons with militarism and a lot of other things that they thought were wrong with the world. The chairman of the meeting, the Dutch physicist Hendrik Casimir, whom I had always considered not only a fine scientist but a good friend, allowed a group of students into one of the scientific sessions, where they denounced the Jasons and, by direct implication, me. Here was a scientific meeting, and those students were let in for a political harangue. I held up my hand and asked to comment and respond, but the chairman refused to let me speak. Hans Bethe, who had good liberal credentials and was no longer directly associated with the Jasons, was allowed to say a few useful things, but overall it was a sort of political ambush.

I felt that the situation was completely out of order. When I got home I wrote Casimir a letter of protest, saying it would kill international scientific meetings to allow politically oriented groups to come in and denounce scientists, especially with no reply permitted in defense. I sent copies to several other organizers of the meeting, too. Well, I got complete apologies, but the incident showed how passionately people felt about international politics of the day, and how people who normally were sensible went overboard and did things that were very inappropriate.

Quite aside from work with the Jasons, during the years (1959–1961) I'd spent at IDA, before most of the problem with Vietnam occurred, it was clear that nuclear disarmament demanded very careful attention. The issues were complex, both politically and technically, and partly for this reason we organized a division of IDA primarily around international political affairs. Some Americans were going outside normal government channels to meet with the Soviets to discuss nuclear problems and related matters. I wanted to take part. While at IDA, I began planning to attend the first Pugwash meeting to be held in the Soviet Union. The Pugwash meetings were initiated largely by Bertrand Russell to discuss such issues, and a meeting of the group in Moscow seemed timely. Of course, Bertrand Russell was not popular in U.S. government circles, and word of my intentions got back to the Pentagon. I soon heard rumblings that it would not look good to have the vice-president of IDA going to a meeting where he might be seen as anything like a peacenik. I spoke with Garry Norton about it and decided it would be wiser to put off such politically touchy meetings until I had left IDA.

In early September 1961, just after leaving IDA, I did in fact go to a Pugwash meeting in Stowe, Vermont, sponsored jointly by the National Academy of Sciences and the American Academy of Arts and Sciences. The Americans there included Jerome Wiesner, Eugene Rabinovitch, Bentley Glass, I. I. Rabi, Bernard Feld, Robert Frost, Paul Doty, and several others—a solid but generally liberal group. The Soviets sent some of their top scientists too, including Nikolai Bogoliubov, Lev Artsimovich, and Igor Tamm, but they ran into a dreadful conundrum.

For the previous several months, the Kremlin had been harshly criticizing the United States for continuing atmospheric testing of nuclear weapons, and had declared a moratorium on such tests itself. The Soviet delegation came over with its speeches all prepared to upbraid the United States about testing. Then just before the meeting, the Soviets lit off an immense nuclear blast in the atmosphere, about 25 megatons, the largest ever. The Russian delegates had not known this would happen and were not sure what to do. I remember well sitting on the side porch of the lodge the first evening of the meeting, while every few minutes groups of two or three Russians came out of the building and headed for the woods to try to figure out what they were supposed to say. Together in the woods, well away from any microphones, they talked over their position. They were somewhat cut off from their government and were in a spot. They tried to finesse the predicament by getting one of their most respected scientists, the chemist Mikhail Mikhailovich Dubinin, to give a long and strong speech against the United States, to deflect the issue. The maneuver did not help much. As I watched Dubinin during his talk and after, it seemed to me he was feeling bad about having to say such outlandish things about the United States, and I felt for him.

The next year there was a still better attended Pugwash meeting in Cambridge, England, sponsored by The Royal Society. This time I had an opportunity to press discussions on making space open to all participants and outlawing attack on space vehicles with Anatoly Blagonravov, head of the Soviet Committee on Exploration and Use of Outer Space, which was a cover for the more powerful but secret organization that really ran their space program. He was a kindly and elderly artillery general who, by virtue of experience with army rockets, had become head of their space committee. He agreed that it was a good idea to forbid attacks on orbiting space vehicles and also to forbid weapons of mass destruction in space. While that was an informal and unofficial discussion, I of course reported his agreement to our own government and have no doubt that he also discussed our meeting with officials in Moscow. Within a year, the United States and the USSR had made a formal agreement. This fortunately was to allow undisturbed satellite observations over both countries, which added substantially to our knowledge of what was going on in the USSR and to global stability.

In the meantime, I had been determined to get back into academic life as soon as the stint at IDA was over. Jim Killian, chairman of the board at MIT, and Julius Stratton, the MIT president, invited me to join MIT as provost. They indicated that they would like to see substantial emphasis on science at the school, already good in science but known best as an engineering powerhouse. I had been approached about administrative jobs at other universities but had never been interested or seen any special reason that I was the person for those jobs. MIT was different. As a science and engineering school, it needed technically trained people in leadership spots, and a boost to its science side would inevitably invigorate its engineering and industrial side too. I felt that it was a place where I might fit, where an administrator with a scientific or technical background was needed. Most scientists are not eager to do administration, but perhaps my ever-present bent for trying something new had its affect on me. So I went to MIT when I left IDA in the fall of 1961.

In general, my experience at MIT was rewarding and the institution was responsive to my efforts. Before long, however, it became clear that some of the changes that I felt to be important elicited very little enthusiasm from industrial and engineering interests. For example, I wanted to draw a line between working in a major way for a private company and being a member of the faculty of MIT. I myself had been offered a position on the board of Perkin Elmer, and I had turned it down. I felt that it was bad form to open the possibility of favoritism toward one industrial concern by a provost who should be properly concerned with all research at the university. I heard later that some of the industry-connected people at MIT were troubled by that. They would have much preferred that I be connected to industry. My position, they felt, implied that I did not care about industry, or that I was even hostile to it. I was an outsider to their culture.

We proposed a rule that nobody on the full-time faculty could simultaneously hold an executive or operating position at an engineering or science-based company, as many had in the past. I felt that a professor in such a position would inevitably find himself putting more energy into his private company, because that would be where his fortune lay, than he would into academic responsibilities. The right and useful arrangement seemed to be to allow a professor to do consulting, or to be on the boards of companies, but not to allow operational positions that would make daily demands on time and energy. The issue went around and around in the academic senate. Eventually the policy was adopted, but not without objections from some of the engineers.

The different interests of scientists and engineers was evident on the national level too. President Kennedy had announced the Apollo (manned lunar landing) Program, an audacious plan hatched mainly by engineers, including German expatriate Werner von Braun. Scientists, and some academic engineers, were rather sour on a manned moon mission. It seemed to me that part of the problem was that scientists had been left out of the decision to try for the moon. They also felt that the program could not possibly have enough scientific value for the costs involved and that it was probably a technological mistake. Of course, scientists felt the moon was a proper topic of study. However, most were convinced that a few automated probes, without people, could do the job much more cheaply than spaceships with people inside.

A groundswell of scientific opposition built up. Alvin Weinberg, a physicist and director of Oak Ridge National Laboratory, wrote an article that said cosmic rays made travel to the moon too dangerous. Jim Killian gave a speech in which he called the Apollo Program an economic disaster for the country because it would soak up so much engineering talent that little would be left to do anything else. Jim Fisk, who had gone to Bell Labs the same year I did and was then its president, said publicly that the estimated $25 billion cost of the Apollo Program was way short. He figured that it would be more like $100 billion and that the project would take much longer than the estimated one decade. And so it went.

One day in 1963 I happened to run into George Mueller, who had just been made head of the Apollo Program. I told him, "You've got all of these people jumping on you and on the program. What you ought to do is bring some of them together and talk with them directly. Have them advise you, rather than talking to the newspapers." I told him that the critics were bright people and that they were serious in their objections. They might not be entirely correct, but they should be listened to. I suggested that an advisory committee be formed of high-level technical people and that some of the objectors be put on it, deliberately, so that this large national program would have the benefit of the best possible advice, both for and against.

Mueller went to James Webb, the head of the National Aeronautics and Space Administration (NASA), and then got back to me: "Jim thinks that's a good idea. We'd like you to form the committee and be chairman."

Things were busy at MIT, but I felt that this was an important and unique program. I tried to get a wide variety of people for the committee who could look at the problems from as many directions as possible—medical people, engineers, astronomers, geologists, physicists, and chemists, including a number of known skeptics. I gave them a set speech when I tried to recruit them, along the lines of, "This is something Kennedy has gotten under way and will pursue. The nation is obviously committed to it. We've got to be talking with them and try to see that the country is realistic and it is done well and properly. Are you willing to be on the committee?" All but one accepted a position on the Apollo Program's new Science and Technology Advisory Committee (STAC). John Pierce, an outstanding electrical engineer I had known at Bell Labs, said no. He turned down the job with the quip, "I've always figured that things that aren't worth doing aren't worth doing well."

We settled into a series of regular meetings. Top representatives of the major contractors were often called in for their input—and also to make sure that industrial participants were adequately informed on the program as a whole and recognized its critical problems. George Mueller always met with us and paid close attention. George also made another committee of the top executives of each company involved. He had them visit one another's plants and explain to one another exactly what their companies were doing, to reinforce feelings of both competition and patriotic teamwork. Stark Draper of MIT headed an important group that was building the inertial guidance system for the Apollo Program in his laboratory. As a consequence, I was asked to be on this executive committee, representing MIT as a contractor.

Thomas Gold, an often brilliant, highly imaginative astrophysicist, presented our advisory committee with one memorable issue. Tommy was convinced we had a serious moon dust problem. He (and a few others at the time) believed that, over billions of years, extremely thick layers of exceedingly fluffy dust had built up on the lunar surface. With no air or wind, but with ultraviolet radiation striking the surface, the dust motes would be ionized, float around, and build up into a uniform but fragile, deep layer, ready to yield at a touch. Gold was a highly persuasive arguer. He gave substantial reasons why the dust would be deep, and he painted a picture of a landing craft touching down, and just sinking out of sight, as dust hundreds of feet thick swallowed it whole. It was an alarming thought. Some of the other scientists on the committee were already skeptical about the Apollo Program, on general principles, and were ready to believe Gold's arguments.

However, there also were strong reasons for doubting the existence of Gold's engulfing moon dust. We asked Dutch-born astronomer Gerard Kuiper, who had been chief scientist on the earlier *Ranger* missions to photograph the moon, to also give our committee a briefing. Kuiper pointed out the very big differences he had recently measured in ultraviolet reflection from different parts of the lunar surface. This indicated that even the topmost layers on the lunar surface differed from one place to another. Kuiper did not have Gold's rhetorical talents; but as a geologist on the committee said, "We'd better keep in mind who is the better arguer here, and be careful." Lincoln Labs had already measured various radio wavelengths reflected off the moon. The results implied that its surface included numerous large, hard chunks on the scale of small rocks to large boulders. Those rocks clearly had not sunk far into any dust. Radio astronomers with the large antenna at Arecibo, Puerto Rico, were also asked to make measurements; these indicated the dust could not be more than a couple of feet thick.

In addition to the swallowed lunar landing scenario, Gold made another anti-Apollo pitch later, again built on worries over dust. He had, as a theorist, thought hard and long about how the lunar surface should behave, in its environment so different from what we encounter on Earth. This second push against the Apollo plan was made in a presentation to a subgroup of the President's Science Advisory Committee about 1968. The committee, he declared, should urge the president to be sure that NASA planned and expected only a single mission to the moon; any more, he felt, would not offer enough additional science to justify the cost or risk of life. His argument was that, since lunar dust would be very mobile—floating around because in the absence of water the particles would be electrostaticly charged—the moon would be covered quite uniformly by dust. Landing at a second position on the moon would give no new information—seen one moon site, seen 'em all. The group's acting chairman, George Kistiakowski, asked for opinions. Essentially all the dozen or so scientists there said, yes, we'd better put a bug in the president's ear to block a second moon mission. As I had been looking at this problem for some time, I felt I had to speak up, and I raised the arguments about radar reflectivity and Kuiper's ultraviolet (UV) measurements. Not only did this evidence suggest much less dust than Gold proposed, it varied from one place to another, implying substantial geologic diversity on the moon. Kistiakowsky was wise enough to realize that there was a valid difference of opinion on the matter. "Maybe we had better wait until after the first landing," he said, "to find out whether Charlie is right before approaching the President."

By and large, I believe the Apollo advisory committee served NASA well and George Mueller did an excellent job of management. The moon landing of course succeeded. Some of the later Apollo Program missions performed first-class geological and other scientific studies. Many of the committee members were skeptical about the whole mission initially, but as they examined the program in detail, and saw that the problems had solutions, they generally became quite positive about it. Mueller really took us into his confidence, always attended our meetings, listened closely to what we had to say, and responded appropriately. The one thing that I have always regretted—and wondered whether we might have done better—concerns the launch pad fire in January 1967 that killed astronauts Virgil Grissom, Edward White, and Roger Chaffee. The capsule had a pure oxygen atmosphere, which NASA changed after the fire. We had never been asked specifically to consider the danger of fire, but we should have thought of it anyway. It was a terrible blow to the program. George Mueller took it particularly hard, and Jim Webb, the NASA chief, worked very hard to help George emotionally through that tragic time. I believe that we did, however, help the program substantially. Besides monitoring it technically, the committee played a significant role in planning scientific experiments. Among other things, we suggested a moving vehicle, a rover, and encouraged putting retro-reflectors on the moon for the laser experiment. And we helped initiate the space shuttle, which has unfortunately turned out to be substantially bigger and more expensive than initially expected.

Figure 13. Meeting of the *Apollo* program's Science and Technology Advisory Committee (STAC) in Houston, Texas, during the first lunar landing on 20 July, 1969. In the far corner of the table is George Mueller, head of the Apollo Project and to his left Charles Townes, chairman of the Committee. The screen at the end of the room is showing the lunar landing in progress.

I of course remember very well the first manned moon landing. Our committee was all together at the Houston Space Center with George Mueller when the landing occurred, and we had an inside view. The astronauts still had to come back safely, but the landing itself was a moment of great triumph and relief.

During those years, I was spending most of my time on academic administration at MIT. I did get some research done, notably in nonlinear optics with two excellent graduate students, Ray Chiao and Elsa Garmire. And Ali Javan, my former student and the inventor of an important type of laser, came from Bell Labs to be a professor at MIT. I not only enjoyed the science, but I was able to get from the students a very different version of life at MIT than what I could get from other professors and administrators—one which was helpful in my administrative judgments. In general, it was a happy time for MIT as an institution. I enjoyed working with Jim Killian and Jay Stratton, who were exceptional administrators and individuals. The administration, faculty, and students were on good terms. Only a few years after I left MIT did the student rebellions of the late 1960s show up—a little retarded at MIT compared to the early 1960s rebellions at the University of California at Berkeley—to create problems.

I took the MIT job because I thought that as provost I could really make a difference, perhaps by strengthening science and, through that, the cutting edge of engineering. There also was a general feeling that I was in line to be president. It was not overtly stated, but it was clear that this would be a natural progression. Well, the dean of engineering, Gordon Brown, clearly did not appreciate all of my moves (though I felt most of the engineers were quite friendly). Vannevar Bush was then still a very important person at MIT, and chairman of the board of trustees. Van had never much liked rockets, and he had objected to my bringing in NASA money to MIT for a space research lab; in his view, it was unethical because the whole space program was wrong and a waste. Van's stance reflected his consistent skepticism about rockets, including whether ICBMs could really work. He also pushed the idea that combustion instabilities would develop in rocket motors big enough for the Apollo Program, resulting in unmanageable irregularities in thrust. At the time, I was on the Apollo committee, and knew that the motors by then had already been designed to be stable.

I clearly had some cards stacked against me. Van Bush was made head of the search committee for the new president. A special faculty meeting was held for the announcement. I walked into the room and several people rushed up to me, calling congratulations, but I knew what had happened and had to say, "Wait a minute. It's not I." It was the dean of the business school, Howard Johnson, who was announced as MIT's new president.

That was a bit of a shock to some at MIT. Now, Johnson was a perfectly good man, but he was not well known in engineering or in science circles. He was, however, a good manager. I too had been surprised when I learned about the decision—but I thought to myself well, okay, they don't want my kind of president, so I'd better get back to research. A little later, Van Bush even apologized to me, saying that he felt bad that an old codger like himself had ruined my career. Clearly he felt responsible, but I never thought he had ruined my career. Instead, perhaps he just helped me get back to the kind of work I enjoyed most. I resigned as provost and was appointed institute professor. Johnson was really very good to me and offered me a fantastic salary, much more than I had gotten as provost, to stay on permanently.

I wanted to take a year off for research, with no administrative responsibilities. That was fine with MIT. I spent a good deal of it over at Harvard, with the astronomers there, looking into areas of astrophysics. I also thought about moving on. I had lots of offers—including a permanent position in astronomy at Harvard, the presidency of Tufts, a deanship or professional position at Chicago—and I had been approached about the presidency of Duke. I had just won the Nobel Prize in physics, which was undoubtedly a factor in all the attention (and more calls from government for advice).

Both IBM and General Electric also asked me to be vice-president for research. In fact, I had considered IBM seriously enough to be interviewed there, where I knew the then head of research, Emanuel (Manny) Piore, quite well. The fact was that industrial work had little appeal to me, so a day later I called them up and said that I really preferred to stay in academic work. The money in industry, of course, was enormously more than what I was getting at MIT or could get anywhere else in academia, so Piore found my decision a bit unbelievable. "Have you talked with your wife about this?" he asked. I told him, politely, that yes, I did talk with her. Piore apologized. Frances was amused by his question but also a little bit insulted by the idea that she would tell me to go after the money.

My primary drive was to get back into fundamental research. While an administrative role was tolerable and I felt could be important, research was more fun for me. If I could not be especially effective in a key administrative role that badly needed a scientist, then administrative work would not be worthwhile for me.

And the Nobel Prize had come my way. This prize, which I received in 1964, jointly with Nikolai Basov and Alexander Prokhorov, provided both opportunities and complications. It was officially awarded for "fundamental work in quantum electronics which led to production of oscillators and amplifiers according to the maser-laser principle," or, as was more simply stated by physicist Bengt Edlén, who officially introduced us at the Nobel ceremony, for "invention of the maser and the laser." This was the same Edlén whom Ike Bowen had invited to Caltech to discuss his identification of puzzling spectral lines in the solar corona and whom I had the pleasure of knowing when I was a student.

Being awarded a Nobel Prize is the most important public recognition a scientist can receive, and it is of course a great moment. Frances and I took our four children to Sweden for the event, and we all enjoyed the rather glitzy, royal occasion, which in the middle of winter and associated with Santa Lucia day, transforms all of Stockholm with a sense of festival. Even Sweden as a whole celebrates and helps make the Nobel Prize what it is. The king, Gustavus VI, was a scholar and an archaeologist—an interesting person. His grandson was being schooled for the succession. The Nobelists that year were an interesting group. Jean-Paul Sartre had been chosen for literature; but he had publicly refused the prize, apparently wanting to make a statement that a small committee was not competent to choose the most outstanding literature of the day. So he didn't come. A couple of years later, when I mentioned to one of the Nobel committee members that I was sorry to have missed the chance of meeting Sartre, he told me, "Well, you know, he did write later and said he would, however, like to have the money!" He didn't get it.

Martin Luther King, Jr., received the 1964 peace prize. I well remember my aunt, Clara Rutledge, telling me of her admiration for the remarkable young minister in her home town of Birmingham, Alabama, who was the then relatively unknown Martin Luther King, Jr. So I was pleased when, as we met, he immediately said, "Are you the nephew of Clara Rutledge? She helped me so much in the early years."

The chemist was Dorothy Hodgkin, an interesting Britisher and only the fifth woman to be so recognized by the Nobel committee in any category of science, Marie Curie and her daughter Irène Joliot-Curie being the first two. The prize in physiology and medicine went to a neighbor at Harvard, Konrad Bloch, and a German, Feodor Lynen. The economics prize had not yet been established in 1964. It is a new one—and not strictly a Nobel Prize—but one established "In honor of Nobel" and given at the same time as his original ones, so that it is hardly distinguishable from the Nobel Prize.

Perhaps particularly for Americans, for whom kings and queens are rather distant, being entertained and feted by royalty is a somewhat out-of-this-world experience. The Nobel occasion is indeed memorable. Yet after a Nobel Prize, scientists can find that their scientific career is more or less finished. The prize often comes rather late in a scientist's career. And Nobelists are then likely to be asked to fulfill a wide variety of public functions, to consult or speak on various public issues, and so can be distracted from the very intense work usually needed for important scientific contributions. Thus while the fame of the Nobel Prize has enjoyable aspects, it can also be a problem. I myself felt that I was still young enough, at 49 years, and eager enough that if I decided to go that way, I could again dive into intense research.

In the research realm, I felt I wanted to move on from laser research and nonlinear optics to fields with more elbow room, where the problems were not yet so clearly recognized, nor avidly followed, by other scientists. I had already left direct maser and laser research behind me and been working on various nonlinear optical effects. But many good scientists were by then in the field. I wanted to work in a field where I could make unique contributions which were being overlooked by other scientists. I decided astrophysics was probably the right place. Astronomy had risen again and again in my mind as an attractive field, a feeling reinforced during my earlier sabbatical leave and by time spent with the Harvard group. It was time to act on those feelings.

California, with its superb observatories and traditions in astronomy, cast a powerful spell. Two particularly attractive centers of excellence were Caltech and the University of California at Berkeley. I strongly considered Caltech, but the top officers there tried too hard to pin me down on exactly

Figure 14. A banquet in the Stockholm Town Hall during the Nobel Prize festivities of 1964. Frances Townes is making a toast. On her right is King Gustavus VI Adolphus, and to his right is Dorothy Hodgkin, winner of the prize in chemistry. Facing Frances and the king is the queen, Louise Mountbatten, and to her right is Konrad Bloch, winner jointly with Feodor Lynen of the prize in medicine.

what I would do. I told the provost that I was interested in interferometry, and in both infrared and microwave astronomy, but did not want to obligate myself to a specifically defined program. In addition, Southern California smog was getting worse, and Frances did not like that. So, Berkeley had the edge on weather.

The Berkeley physics chairman, Burt Moyer, called and invited me up to talk with the president, Clark Kerr. I suppose Berkeley had been my favorite from the start. I had had an opportunity to consider going there once before, shortly after I had gone to Columbia, but at that time had not wanted to make another change so immediately.

Politically, Berkeley was a lively place, with the Free Speech Movement, the antiwar protests, and other such activities. Many people who had lived in Berkeley for years were moving out of town, heading over the hills to more peaceful towns to the east. They believed the city was falling apart and would not be a good place to live. I thought they misjudged the situation and did not worry much about Berkeley's political atmosphere. What did concern me was how California's new Republican governor, Ronald Reagan, was going to treat the university. There he was, a tough self-styled conservative, while Berkeley, with all its political radicalism, was the leading example of a lot of things he most opposed.

I met Kerr in his office and raised the Reagan question. He explained, very civilly, that Reagan was not the first governor to come into office with a low opinion of the university. Pat Brown, Reagan's predecessor, initially had doubts about U.C. but, after getting to know it better, was one of its great supporters. The same thing, Kerr assured me, would be true of Reagan. I took the job. However, Kerr's political judgment was a little off. Not long after I took the job, Reagan dismissed Kerr (who, in his most quoted remark, said that he left the same way he came in: "Fired with enthusiasm"). Nonetheless, I have never regretted the decision to move to Berkeley.

A few of my friends were amazed. After I had stepped down as MIT provost, I had accepted the invitation to go on the board of the Perkin Elmer Corporation. Admiral Chester Nimitz, the company's chairman of the board, was thunderstruck when I told him where I was going. "Charlie, how could you move to Berkeley? That's the most sinful city in the whole United States!" Of course, the by-then notorious culture in Berkeley held business, government, and the military all in equally low regard.

What had been offered to me was a special position, called professor-at-large (now called a university professor). I was also offered a secretary and $100,000 to build up a laboratory, with no significant restrictions on how I used it. My office in Birge Hall looked out at the Campanile, Berkeley's landmark tower. Later, as a Nobel laureate, I got one of the campus's most treasured perks: a reserved parking spot. All that was hard to beat.

Before accepting the job, I called another professor-at-large, Harold Urey, at U.C. San Diego, to ask just what the obligations were for a professor-at-large. The title meant, officially, that I was attached somehow to all the campuses and reported to the U.C. president, rather than to a departmental chairman. Did I have to move around or what? I asked him if he really reported to the president, and how. "Oh, yes, I report to and advise the president," he said. "Why just last month I wrote him a letter about not allowing some trees on the campus to be cut." That type of function certainly did not seem onerous. So it really was an ideal position, with much freedom and primarily the duty to do good work. Of course, I agreed to speak on the other campuses frequently, but Frances and I settled comfortably into Berkeley and bought a home north of campus in the lower Berkeley hills. Our daughter Ellen had also independently just decided to go to graduate school in biology at Berkeley.

The people in the department made us feel right at home. One reason may be that Berkeley was feeling a little beleaguered at the time. It was a positive thing to show that they could bring in a new person. The politics of the day did not interfere too much with the physics department. We occasionally had tear gas on campus, and I would watch the activity and commotion out the window, but unlike other departments, where riots and disturbances did affect the scholarship level badly, most of the physics students stayed busy with their work.

Still, it was Berkeley. I had remained active with the Jasons, which entailed occasional work with classified material. I brought with me a safe, to store sensitive papers, and put it in my office, where it remains today. There was no other such safe on the campus. I of course checked with Roger Heyns, the chancellor, to be certain the safe for classified documents was not against official policy or would not be too much of an incitement for radical political attacks. I told Heyns that doing the work related to the safe was a matter of principle with me—that one should not give up on government as long as it is listening to one's advice and as long as the work seems useful. He agreed.

Some time later, I received a call from the campus police, who asked, "Do you realize you are on the front page of the *Berkeley Barb* as 'Dr. Strangelove'?". The *Barb*, a counterculture weekly, had a picture of me accompanying an assertion that I kept antipersonnel weapons in the safe. The campus police were worried about my safety and wanted to change my locks and post a guard near my office door. I told them to just hold on. My graduate students had keys to my office and I wanted them to come in and out freely; the whole atmosphere would change if I had special locks and a guard. I read the article and decided the danger was probably small. The police reluctantly agreed to leave it up to me—and there never was any serious problem.

One of the university's most dedicated antiwar activists was physics professor Charles Schwartz. Charlie had a number of followers among the physics students, including some students with whom I worked. I liked them, and I think they liked me, but they did have intense, emotional feelings against doing anything that might support the government or the military.

One day, during the lunch hour, my secretary came back to my office early and found Charlie going through my files, apparently looking for incriminating evidence that would paint me as some sort of warmonger. I would not have done anything about it if I had caught him, but she reported him to the chairman of the department. He insisted that Charlie write me an apology, which he did, in a rather short note. Of course, he could not get into the safe. My biggest worry was that he would mix up my files. Anybody could look at those files—all he had to do was ask.

In 1967, as president of the American Physical Society, I had invited President Johnson to speak at a Washington banquet during the society's annual meeting. It was the time of the Vietnam war, and as the President was about to speak, a member of the society rose from one of the banquet tables waving a placard. He tried to say something, objecting to White House policy, but others at the table sort of pushed him back down.

That meeting had a repercussion a few years later in Berkeley when I was asked to come listen to, and comment upon, a speech by Washington journalist Dan Greenberg about science and the national scene. After his speech I made a few comments but was interrupted by a flurry of questions from the audience about that Washington banquet, where Johnson had spoken, and by accusations that I had done something terrible by inviting him to the American Physical Society meeting. I explained that I felt the president of the United States had a pretty important role in the country and that it was a good thing for physicists to get acquainted with him and his views. Again, Charlie Schwartz was critical—and was particularly vehement about how I had kowtowed to government.

After this meeting, Stephen Smale, one of the meeting's organizers and an outstanding mathematician, invited me over to his house and said he presumed I realized that the affair had been planned as a chance to attack me. He was very friendly and said my responses had persuaded him that what I was doing was honorable after all. Charlie Schwartz, however, never could accept any of those connections with government, business, or the military as ethical behavior by a university professor. He continued to look for the errors in my ways and from time to time attacked me in the student newspaper or during meetings of the faculty senate.

I have always felt that independence of mind and thought are very important, not just in science but for civilization in general, so I never objected greatly to the debates at Berkeley over policy. In fact, I agreed with a great deal of what the antiwar people were saying, but I could not go along with just being arbitrarily against government or the military. The Vietnam war was a policy mistake, and what was being done was poorly carried out in my view, which made it all the more important to stay in touch with the policymakers and try to advise them as well as one could.

During those years I was very sympathetic with the young people's problems with the war. A number of the grants to my lab came from the military, particularly the Navy, which I have always considered to have excellent policies for the support of research. I told my students, "This money is completely open. We use it the way we want to use it. Are you sure you feel comfortable about doing that?" We were, after all, simply doing the physics we found interesting—spectroscopy, quantum electronics, and astrophysics. Some Berkeley students strongly opposed military contracts on principle, but there were dilemmas and inconsistencies. Charlie Schwartz, while he spoke out vigorously against using any military money, actually had received Air Force support for some of his own research. As it happened, even in the tensest of times on campus, most of my students seemed to have no trouble with research support from the military, but some appreciated the idea that they could make a choice.

It was in this tumultuous setting that, in 1971, I received a wholly unexpected call from James Roche, chairman of the board at General Motors. He said he was coming to the Bay Area and wanted to have dinner with me. I had never had anything to do with automobiles, except to drive them, so the call seemed very puzzling. We met at a restaurant near the Oakland airport. He wanted me to set up a technical advisory committee for GM. The company was being attacked severely on safety, pollution, and other fronts, and it had come around to realize that it needed outside help.

At the time, GM accounted for about 3 percent of the gross national product. It was important to the nation. Roche agreed to all my conditions—that I have essentially free rein over choosing the members of the committee, that we report directly to GM's executive committee, and that I have final control over any press releases about our activities. This last point seemed important to me because I did not want to be part of an empty public relations exercise.

Today, if a major corporation asks a university professor to help it out, nobody raises an eyebrow and the university community is generally pleased with such a connection. Back then, especially at Berkeley, big business had a reputation on campus that was nearly as low as the Pentagon's. I asked U.C. President Charles Hitch if he thought that working for GM in this way was acceptable. I knew it would cause a stir on campus. If he had said no, then I would have turned it down, but he listened to my argument that since GM is so important to the country, it was sensible and a useful thing to do. So, after a little thought, he said, "On balance, I think you ought to do it."

I was very careful to get good people with a variety of pertinent backgrounds. The committee included Lee Du Bridge, physicist and president of Caltech; Bob Sproull, physicist and president of the University of Rochester; Martin Goland, mechanical engineer and president of the Southwest Research Institute; Bob Morison, biologist at Cornell; Bob Cannon, aeronautical engineer at Stanford; and Ray Baddour, chemical engineer at MIT. We made a rule that meeting dates would be picked when everyone could attend—and they did. The committee really worked hard, going to visit factories and meeting with GM's design and manufacturing division chiefs, as well as top executives. The GM top officers listened to us closely. We even warned in our first year's report that Japan was poised to provide very stiff competition—that pollution and safety problems could be solved, but Japan could present a more difficult long-range problem. That advice now seems obvious, but it was somewhat new then. One year our report came down a bit hard on GM's quality-control efforts, calling them insufficient. I don't think they took our committee very seriously on this point, but they did respond favorably to enlarging GM's research contingent. I hope we gave them some help, but it was tough for management to readjust such a large corporation to changing times and growing Japanese competition. After three years of this, I said that I should resign, feeling the essence of our job was to provide a fresh view and, hence, a fresh chairman was needed. GM then asked me to serve on the board—I got kicked upstairs. This was the second industrial board I had agreed to serve on, and I always felt that two was the limit for me. Such contacts with industry were informative and interesting, but more would leave too little time for my real work in physics. As time went on, the Berkeley campus became pleased with my industrial contacts, rather than shocked or hostile, as it had been in the late '60s and early '70s.

At Berkeley, I am like the middle of a political sandwich. Many members of the university community regard me as excessively affiliated with big business and government, and generally pretty conservative, while business people often think of me as a Berkeley liberal and perhaps just a little bit dangerous.

I have continued to spend regular stints on various government policy committees. One of the most challenging and arduous of these came in 1981, early in the Reagan administration. I had gotten acquainted with Caspar Weinberger some years earlier, during meetings of the Bohemian Club, which gathers every summer in a redwood grove near the Russian River in Sonoma County, northern California. Weinberger and I were members of the same subgroup at these informal meetings and got to know each other fairly well. After Reagan named Weinberger his secretary of defense, Cap called to say that he wanted me to chair a committee on MX basing.

The MX was an intercontinental ballistic missile with multiple warheads. The Carter administration had begun a program to build 200 such missiles, and it had intended to use a deceptive mode of basing them: scattered across Utah and Nevada would be ten silos for each missile, or 2,000 holes in all. Railroad cars would shuttle them here and there, under cover, at irregular intervals. The idea was that the sheer number of targets would make it impossible for the Soviets to mount a surprise and crippling "first strike" attack which would be able to destroy them all.

The exact manner of deployment had been left unsettled when the new administration entered the White House. So many billions of dollars were at stake that Reagan's advisers wanted a new look at the problem. The MX was relatively small, as intercontinental missiles go, precisely to permit mobile or flexible basing. It could fit in a railroad boxcar, a large truck, or in the bay of a transport plane. A number of possible schemes presented themselves, the primary one proposed by the Air Force being the original "racetrack" idea from the Carter administration.

Among others, I consulted an old friend, Spurgeon Keeney, for advice on whom to appoint to the committee. Weinberger probably would have disapproved my calling on Keeney, as he was fairly liberal and had been close to the Carter White House, but I trusted him. He knew the system and most of the important players in Washington. The final committee included Admiral Worth Bagley, retired from the U.S. Navy; Dr. Solomon Buchsbaum, physicist and senior vice-president of Bell Labs; General Andrew Goodpaster, former supreme allied commander in Europe; David Packard, chairman of Hewlett-Packard; Professor William Nierenberg, a physicist and head of the Scripps Oceanographic Institute; Dr. Henry Rowen, a professor at Stanford and chairman of the National Intelligence Board; former Air Force General Brent Scowcroft; General Bernard Shriever, former commander of the Air Force Systems Command; Dr. Albert ("Bud") Wheelon, head of Hughes Aircraft Laboratory; and the Hon. R. James Woolsey, former undersecretary of the Navy. Invited observers who met regularly with us and provided important discussion were the Hon. Stephen Ailes, former secretary of the Army; Lieutenant General Glenn Kent, former assistant chief of staff of the Air Force; and Dr. Michael May, associate director and former director of Lawrence Livermore National Laboratory. It was a powerful group, and the problem was important.

Weinberger was clearly interested in our best advice, not just a cover rationale for any plan already favored by the White House. "Tell us what the technical situation is," he said. "Leave the politics to me and the President." I asked him how the final decision would be made. He said Reagan would make the call and needed a detailed presentation. I was surprised. The impression one got in those days was that Reagan was not one to get involved in detailed decision making. But that was not how Weinberger described the way the White House worked. He clearly regarded Reagan as very much in control on matters of this type.

We also had a meeting with several of the senators from the Western states most likely to get the bases. They told us that, frankly, their constituents didn't want the missiles. But, they said, if we felt that the country really needed the MX in their states, they would support our position.

In our deliberations, we noticed a pronounced contrast in the quality of advice we got from employees of the national laboratories, such as Livermore or Los Alamos, compared to those from industry. Boeing Aerospace, for instance, perhaps stood to gain most from the multiple-basing ideas and came up with plenty of arguments for them. The national lab people seemed very professional, well informed, and open to varying opinions.

In the end, we decided against any extraordinary effort at deceptive basing, particularly the 2,000-hole idea, because that could not safeguard the missiles adequately. It appeared to us that no matter how much money we spent, the Soviets could find a way to make a first strike plausible and could probably spend much less money than we would. In fact, all the elaborate basing schemes seemed to have technical or political problems, or both.

Senator John Glenn pushed us on one particular idea: to put the missiles on trucks that would continually move around on highways and thus be hard to hit. Technically, that seemed possible, but we felt that the idea of putting them on public roads spelled certain public outrage. We could just imagine protesters out blocking the road whenever a suspected MX missile van was spotted. The committee's final advice was to build one silo for each missile, a conservative and relatively affordable system. We also concluded that each missile might possibly be carried by an airplane. With enough planes out over the ocean at any one time, in unknown and rapidly changing locations, they could safely provide a counter strike. This too could raise public objections to the idea of having a nuclear missile cruising overhead, even if most were at sea, but it presented less of a problem than Senator Glenn's proposal. There seemed to be no really happy solutions.

My personal stand, supported by only two other members of the committee, Bill Nierenberg and Admiral Bagley, was not to deploy the missiles at all. We put this as a minority view in our report. I felt that to build a few missiles, and demonstrate the industrial capacity to build many more in a relatively short period of time, would put enough pressure on the Soviets to agree with us to avoid more multiple warhead missiles. For me, that was the best immediate plan. Nierenberg tends to be conservative on defense issues, but he later told me he was very pleased to take a position against what could have been a significant escalation in the arms race.

Weinberger asked me to make a personal presentation to the president. Before this, Cap took me aside and asked for my personal views. Of course, I urged a minimal deployment and elaborated on why.

My meeting with Reagan was, if I remember correctly, at a special setup in a hotel in the Los Angeles area. I presented the majority report—essentially to build 200 silos for 200 missiles—and also our minority report, not to deploy any missiles initially. General Lew Allen, head of the Air Force, was there to defend the 2,000-hole strategy. The pertinent cabinet members were present. William Casey, director of the CIA, asked a lot of questions and seemed to take the 200-silo approach as a serious possibility, while former General and then Secretary of State Alexander Haig was pretty much for Lew Allen's 2,000-silo position.

I got a funny impression of the president. His behavior illustrated why the public had such a mixed view of him as a combination of a very persuasive and firm leader but a man who sometimes seemed not to be fully aware of things around him.

After we all gathered, Reagan came in. Then, as we were talking and reviewing the options, and also during my own presentation, Reagan very slowly took care of his contact lenses. He took a long time with them, all while we were discussing these very important matters. He had a glass of water and washed the lenses, put them on, took them out, and so on. I thought to myself, "My goodness, how much of this is he going to understand?" But toward the end of the presentations, he asked questions of his own and they were, in fact, very perceptive.

It did not take Reagan long to make what seemed to me a very sensible decision: to build just 40 MX missiles and put them in one silo apiece. I was very pleased with this, as it went a long way in the direction of the minority report and seemed to me enough to impress on the Soviets our ability to build these things, without deploying them in unnecessarily large numbers or being extremely threatening.

During my time with committees such as this, I always made it a rule to keep relevant matters confidential until our reports were made public by those to whom we gave them. Yet keeping secrets in government is not always easy, especially from the inquisitive and resourceful press. Let me give a few examples here.

In the early 1970s I had been chairman of the Space Sciences Board of the National Academy of Science, which was advising NASA on its agenda for space science and solar system exploration. One day I got a call from a reporter at *Time* Magazine's Los Angeles bureau. He asked why the board opposed the "Grand Tour," a proposed mission that would have sent a single spacecraft on a long and complex mission to visit all the outer planets, from Jupiter to Pluto. It would have taken clever advantage of the gravitational field of each world to deflect the trajectory to the next planet on schedule; but it would have been costly and, we felt, much too complex and broad in its aims. Money might be spent better on more targeted missions, for example most immediately on a mission to Jupiter.

Well, our discussions were supposed to have been confidential, but a reporter had been leaked the information and I had a pretty good idea of the general source. The Grand Tour was an idea of the Jet Propulsion Laboratory (JPL) in Pasadena. People at JPL were pretty caught up in it, as though JPL's future depended on this one mission. The committee included two people from Caltech, which is closely affiliated with JPL: physicist Willie Fowler and biologist Norman Horowitz, both of whom had been at Caltech with me when I was a student. I called them. Willie had no idea how the reporter had gotten his information, but Norman said it was probably his fault, since he had told one of JPL's chief proponents of the Grand Tour, Bruce Murray, that we had recommended against it.

Norman said that he had no idea Bruce would go to the press. The affair shows how easily word gets out. About that time, Bruce called me himself and let me have it. I told him I was sorry that the committee was not enthusiastic for the idea, but it wasn't and that I shared its skepticism. Bruce said he was going to continue to take the issue to the American people and probably would have, if JPL's director Bill Pickering had not talked him out of it. In the meantime, the reporter's version of our position was just enough off base so that I was able to tell him that he did not have the correct story and that I could say no more about it. The issue died down until our recommendations were released formally. As it turned out, JPL then planned an excellent but much simpler expedition by the *Voyager* spacecraft to Jupiter, more or less in accordance with our recommendation. Ingenious JPL engineers were then able to command it on to the outer planets, much as had been hoped for from the Grand Tour. Overall, it was a great success—and Bruce Murray, I hope, was pleased.

Another time, during the IDA years, columnist Joseph Alsop took me to lunch and asked me for information about a report on missiles. I wouldn't tell him, and he launched into a long lecture about the right of the American people to know and my duty as a citizen to tell him. It was quite a marvelous lecture, but I did not give him the information.

One day a reporter called and asked about a similar matter, but I told him that because the report was to the White House, only the White House could release the information. Three hours later the reporter called back and said the White House press secretary had given the okay for me to talk. I replied that as soon as I got word directly from the White House, I'd open up. Of course, I never did. Clearly the reporter was lying. It means so much to the press to get this kind of information and beat their competition that reporters will try all sorts of things. Often, their schemes work.

Almost an entire letter I gave in confidence to Defense Secretary Casper Weinberger wound up, within a week, in the hands of the *Washington Post*. As a follow-on to the MX committee, I was asked to head a committee on a subsequent Air Force proposal, the so-called dense pack scheme. In this scheme, many missiles were to be based very close together, which would make protection a little easier and sudden destruction of all of them more difficult. Our committee report was somewhat bland. Weinberger then asked me for my personal views, which I wrote out in a letter to him, expressing considerable reservations about the dense pack scheme. The newspaper's version of my letter was correct almost word for word. The text had probably been read to a reporter over the phone, so fortunately there were a few errors. This allowed me to say to the reporter "No, this is not accurate and that's all I can say." Weinberger was furious. He never did find out how it got out, but he had shared it only with the Joint Chiefs of Staff. The Navy particularly did not like the MX and may have been the source of the leak, but I do not know.

I have seen a few people brought in as advisers who then undercut their influence with government by going public with their feelings. For instance, my service on the President's Science Advisory Committee extended into the Nixon administration. A hot issue at the time was whether to build a supersonic transport (SST) to compete with and surpass the British and French Concorde program. Our committee generally opposed the idea, and eventually Nixon did cancel it. In the meantime a member of his advisory committee, Dick Garwin, was questioned by a Congressional committee and testified against it before Nixon had announced his decision. In his testimony, Dick avoided giving any information restricted to PSAC insiders, but Congress was well aware that he was a member of the team. I think one reason Nixon became more and more distant from his science advisory committee was due to events such as this, as well as to differences in views about Vietnam. He felt its members were not a completely trustworthy part of his team—and PSAC was in fact shut down during his administration.

All this is simply to point out that the pressures in Washington to talk out of school are enormous, and leaks are impossible to avoid entirely. Nonetheless, I have tried hard, when asked to give private advice, to do everything I can to keep it that way. Such efforts may be one reason that I was called on frequently for advice by Republican administrations, particularly Nixon's and Reagan's. Then, too, with both Nixon and Reagan, I tried to press them at times in directions that were somewhat against their philosophy. So while they asked me for advice early in their administrations, they didn't keep asking me back.

Among the most furiously argued initiatives in military technology that the United States has ever pursued was the Strategic Defense Initiative (SDI), better known in public discussion as Star Wars. It was an immensely audacious idea, to knock out enemy ballistic missiles with antimissile missiles, extremely powerful laser beams, or perhaps even beams of atomic particles.

My first hint of it came in March 1983, with a phone call from Sol Buchsbaum, an excellent scientist and friend who was calling on behalf of Jay Keyworth, the White House Science Adviser. I was told President Reagan was going to make an important speech at the White House and that I would find it very interesting, so would I please come, with my travel paid?

Even though they would give me no hint of what it was about, I went to the White House and listened. The president explained it all rather eloquently, invoking a plan to end the balance of terror, also known as the doctrine of "mutual assured destruction," or MAD. No longer would the American people be helpless in the face of missile attack. He would even consider sharing the technology with the Soviet Union, to make the security of each nation based more on the power of defensive weapons than the threat of mass killing.

It was an idealistic plan, and one that would keep the technical community busy. But I immediately had doubts. It would be no easy thing to aim defensive weapons at perhaps hundreds of incoming missiles, very likely coming in a surprise attack, pick them out from possible swarms of decoys amid an electronic clutter of enemy broadcasts designed to befuddle our command and control system, and destroy essentially all the rockets in 15 minutes or less.

I was told that, well, yes, it had some difficulties. But the president's science adviser's office had reviewed Reagan's proposed speech very carefully, and it contained nothing technically incorrect. In fact, this speech was defensible. In my view, some of the many things said later were not defensible, but I believe that first speech contained nothing that was technically erroneous.

That said, would it really work? And how did it come up? It was all Reagan's idea, I was assured. I asked Buchsbaum, did he really think it would work. He got a bit cagey at that point. The high military and White House staff were of course there. General John Vessey, chairman of the Joint Chiefs of Staff, told me, "You know, it is kind of funny. We are not all that certain how well it could work, but the president wanted to announce this policy." But, again, how did it all come up? Vessey said, "The president called us up and said he would like to come over and see us." This, Vessey said, was rather unusual. "We thought, well what kind of discussion can we have with the president that will interest him?"

For several years, since the Carter administration, in fact, the Defense Advanced Research Projects Agency (DARPA) had pursued a quiet examination of ways to knock out ballistic missiles. A few outsiders, such as Senator Malcolm Wallop, were enthusiastic over the effort, but few people thought it was ready for full-scale development. The Pentagon generally supported it, just in case it might be useful, or that the Soviets were working on something similar. The military did not want to get caught by surprise, so they pumped a steady supply of money into it.

On the occasion of the president's visit, the Joint Chiefs of Staff decided to give him a discussion about the desirability and perhaps long-range possibility of complete defense, apparently regarding it primarily as a somewhat philosophical discussion. When Reagan went over to the Pentagon and heard a good presentation of the idea, he loved it. It was just exactly what he wanted to do: find a way to break the nuclear stalemate and end the constant terror of nuclear Armageddon that had hung over the American people for nearly a quarter of a century. In essence, the Pentagon shared with the president an idealistic and speculative missile defense, and he latched onto it.

Judge William Clark, who was then U.S. national security adviser, and Robert McFarland were at the meeting and told me that, yes, there were some reservations about its practicality, but nobody wanted to disappoint the president. There seemed also to be the feeling that, as Cap Weinberger commented to me later, "When scientists get going, scientists can always do these things." There was a faith, in the wake of the Manhattan Project and the Apollo Program, that with enough money and effort, scientists could overcome any obstacle.

Edward Teller was at that meeting to listen to the speech, too. Today, many people believe that Teller was the man who sold Reagan on Star Wars, but that is not the real case. I asked Teller directly at the White House that day if he had been talking to the president about it. He replied, "I haven't met with the President in quite a while now." It is clear that Teller did like the idea, later pushed it enthusiastically, and continued to do so long after evidence accumulated that it was unlikely to work—but it was that briefing by the Joint Chiefs that was the spark for Reagan's enthusiasm.

The problem was that Star Wars was not only an immense engineering job, it had a moving target. The Russians could modify their missiles and other offensive weapons faster than we could ever build leak-proof defenses. I questioned it when the floor was open for discussion immediately after Reagan's talk. I spoke with Cap Weinberger about it a number of times. Yet the die was cast—it was the president's decision and was to proceed. My old friend the laser was a central theme of the program—powerful lasers, perhaps X-ray lasers—to almost instantaneously knock down incoming warheads. Still, I had to generally oppose the idea.

I firmly believe that the success of the Apollo Program had something to do with the adamant support for the Star Wars program by Washington officials. In the Apollo case too, scientists had come out strongly against the practicality of sending astronauts to the moon. Yet, with money and hard work, it succeeded. There was some confidence that also with Star

Wars, given enough effort, a way would be found. I myself felt that some substantial effort was warranted, because we needed to look carefully at the possibility, but that the moving-target aspect of the problem made it quite different from the moon program. The attacker would always hold the advantage. Eventually, the program was to decay into a more modest effort, with much more limited goals.

9

THE RAINS OF ORION

Ancient Masers and Lasers

During a 1963 meeting in Washington, D.C., a group of radio astronomers had dinner at the home of Gart Westerhout, a Dutch scientist then on the faculty of the University of Maryland. Alan Barrett of MIT, one of the guests, soon had the party buzzing over his news. He and colleagues from Lincoln Lab had detected a signal from space carrying the signature of OH, the very reactive, or "free radical" hydroxyl molecule. This was the first molecule discovered by radio astronomy, and only the fourth known in interstellar space (after CN, the cyanide radical, and CH and CH+, methine radicals).

Harold Weaver of the University of California at Berkeley listened closely. First thing the next morning, he alerted the crew at Hat Creek, a U.C. radio astronomy observatory in a mountain meadow near Mount Lassen in far northern California. Within days, the Hat Creek team confirmed that, yes, there it was, the distinctive spectral line of OH. Almost at the same time, astronomers in Australia and at the U.S. Air Force's Cambridge Research Labs also confirmed the result.

Before long, we held a press conference at MIT to celebrate Alan Barrett's discovery. As provost, I introduced the topic—discovery of OH—and told the reporters that this was very important news. Perhaps, I told them, it was not as crucial as the earlier discovery of the 21-centimeter line of hydrogen gas in space used to map the spiral arms of the galaxy, but it was a significant addition to our knowledge of interstellar space. Nowadays, I put it right up there with the 21-centimeter line. The OH signal was to open our eyes to a previously undetected phenomenon among the stars—one of great personal interest to me, in addition to its scientific value.

During the 7 years that I was primarily occupied with science policy advice and administration, in Washington and later as provost at MIT, I never doubted that my first interest was research. There is always another stone to turn over, to see what discoveries lie underneath. The lure of astronomy, in particular, had attracted me at intervals during my years at Caltech, Bell Labs, and Columbia, and it was still strong.

Even before the full import of Barrett's discovery became clear, it hit home with me. I knew Alan well. He had been a student of mine, from 1953 to 1956 at Columbia, and did a thesis in microwave spectroscopy of molecules. He was also a paragon of persistence. He finished up at Columbia just after the time I had discussed—during my 1955–1956 sabbatical and at an international meeting in Great Britain—the good chance that many molecules may exist in the hydrogen-rich and helium-rich nebulae among stars, and that they might be detectable at microwave frequencies. We had also succeeded, just before I went on sabbatical, in measuring some microwave spectral lines of OH at Columbia, and hence had a reasonably good handle on its frequencies. Hydroxyl clearly was a prime candidate, along with others such as carbon monoxide (CO) and my old friend ammonia (NH_3). After finishing his Ph.D., the detection of OH in space became Barrett's consuming goal.

Alan and another new Ph.D., Edward Lilley, went down to the Naval Research Laboratory in Washington and tried, without success, to find OH by tuning to the expected frequency with NRL's large radio telescope pointed at likely looking nebulae. Ed moved on to Harvard and other pursuits, while Alan joined the astronomy department at the University of Michigan and kept up his interest in OH. I remember a call I got during those years from Leo Goldberg, then the Michigan astronomy department chairman. Leo was worried. Alan, he fretted, wasn't quite working out. "You recommended him to us, but the only thing he seems to want to do is this OH. But is that ever going to get anywhere? . . . He's already failed once." I said that it seemed worth a try, and I had confidence in Alan's competence. Leo was not fully convinced.

To undertake a laborious experiment that others think will not work is a courageous and risky thing for a postdoc to undertake. If nothing else, lack of success can make job opportunities scarce. Alan, with his search for OH still unrequited, was lucky to be able to go from Michigan to the electrical engineering department at MIT.

In the meantime, my group at Columbia had measured with some precision the lowest frequency OH absorption lines, at 1167.34 and 1665.46

megahertz, or at a wavelength of about 18-centimeter. These frequencies were also the ones to be most expected in interstellar space.

At MIT Alan linked up with Sandy Weinreb and some Lincoln Lab engineers to use a new receiving technique involving autocorrelation, a method made practical as computers were developed with increasing power. That detector technique, coupled with knowledge of the precise frequency, did the trick. After his wander in the wilderness, Barrett's career was suddenly bright and safe. What had seemed a stubborn preoccupation to some could now be reinterpreted as proper persistence and clear, steady vision.

New discoveries often come with a fresh supply of puzzles. This was no exception. While Weaver's Hat Creek team in California confirmed Barrett's basic discovery, it was nonetheless baffled. One morning, at about 3 A.M., Harold got a call in Berkeley from a technician at Hat Creek. The antenna had been aimed at a bright nebulosity. The astronomers had expected an absorption dip, or line, owing to the interception of the background radiation by OH at its characteristic wavelengths. "There's something wrong," the technician told Harold that morning. "I just can't figure it out. Everything seems to be working all right; we've got some signals at the right frequency, but of the wrong sign—the power goes up instead of down. Could you suggest what I might be doing wrong?"

Harold and his colleagues, with an antenna superior to what Barrett had at Lincoln Lab, scrutinized many nebulae where OH was emitting rather than absorbing energy. They already knew that the OH spectral signature exhibits what spectroscopists call hyperfine structure, represented by a set of closely spread frequencies. By 1965, the California group had established that these often came with strange, unexpected relative strengths of the closely spaced lines. Some, as had been already reported by that technician, were remarkably bright, as though an immense transmitter was beaming powerful signals across the galaxy. Harold saw that there clearly were processes under way out there that could not be explained in terms of dust and gas glowing with ordinary thermal excitation. And not just a few places had these bright interstellar lines in them. In parallel with Weaver's group, John Bolton and his associates in Australia had also noted the anomalous intensities in OH spectra.

Harold began to wonder if the source could really be OH—and not some strange new substance. With a bit of flair for the dramatic, he devised the term "mysterium" for it. His coinage was an echo of the discovery of helium, first detected from never-before-seen features in the sun's visible spectrum in 1868, before the element was found on Earth in 1895. It also recalled a "discovery" in 1865 of a newly suspected element called "nebulium," from a set of green spectral lines in nebulae that corresponded to nothing observed in laboratory studies. (In 1927, Ike Bowen demonstrated that nebulium's signature actually was due to unusual emissions—called "forbidden lines," that arise from oxygen and other well-known atoms under conditions of very low densities and high degrees of ionization—which had not been seen in the laboratory.)

I had followed the mysterium development with some interest, but no real analysis, since I was rather preoccupied at MIT. However, the formidable Russian theorist Iosef Shklovskii gave it serious thought. His interest was natural, for he and I, independently, had been the first to predict that astronomical OH might be detected at microwave frequencies. I remember his bringing up the subject when I met him on my first trip to the Soviet Union, in 1965, and his suggestion that the mysterium signals were probably emitted by natural masers in space.

Masers in space! It clicked with me immediately. This did seem likely—and a delightful thing to contemplate. Out in space are tremendous regions of gas, mostly atomic hydrogen but also various molecules, pumped out of thermal equilibrium by radiations from stars and by atomic collisions. These are very natural places for nonequilibrium. Of course, natural molecular masers in space could have no resonant cavities. They would be real masers, however, in which electromagnetic waves coursing through nebulae pick up energy via the very same mechanism of inverted energy levels and stimulated emission as did laboratory masers and lasers.

The maser model was clinched by Bernard Burke, Alan Rogers, Alan Barrett, and Jim Moran at MIT. They ganged together two radio antennas at Lincoln Laboratory's Haystack Observatory to make an interferometer able to obtain very high angular resolution on the OH sources. They showed that the powerful emanations from OH sources were emerging from such small volumes in space—too small to be resolved by the telescopes—that only masers made sense. Such energies from a mere thermal or blackbody process would require temperatures too high for OH molecules to survive.

When I arrived in Berkeley in 1967—determined to move toward astronomy in the infrared and radio wave regions of the electromagnetic spectrum—microwave spectroscopy of the sky was a natural starter for me. Despite Barrett's exciting discovery of OH, since then no other molecules had been detected. The general feeling among astronomers seemed to be that OH and the other three free radicals that had been found in space, CN, CH, and CH^+, were all there was. The idea was that gas densities are so low in nebulae, and ultraviolet radiation so intense, that any normal molecules surviving would be too scarce to be detectable. No one seemed to be taking seriously my discussion in 1955 of the variety of molecules that might be detected. I figured it merited at least a look. In addition, my reading about radio measurements of hydrogen atoms in interstellar clouds had turned up a couple of discussions about interstellar dust clouds in which no atomic hydrogen was seen at all. It was amazing that the clouds contained dust particles, but no atomic materials. There was a little speculation that in those peculiar cases the hydrogen was molecular—that two or more hydrogen atoms were combined—but that seemed not to be taken seriously by most of the astronomical community. However, Tommy Gold and Fritz Zwicky, two astronomers who had produced many radical but interesting ideas, both made the claim, in 1961, that molecular hydrogen was probably common. My thought was that if indeed hydrogen was forming molecules in space, why not other combinations of atoms? The first molecular candidates I had in mind included ammonia, carbon monoxide, hydrogen cyanide, and a few others.

Jack Welch, an important figure in the radio astronomy group at Berkeley, suggested the use of a 20-foot antenna at Hat Creek, which was just being completed, and said he'd help me get started. It would be the new telescope's first project. Despite his particular interest and generosity, I got the feeling that most of the Berkeley astronomers thought my idea was a little wild.

I was not completely alone, however. Jack Welch told me later that he himself had talked about using the new antenna to search for some of the molecules suggested in my 1955 talk, but the idea had been squelched by other astronomers as a waste of time. At about the same time, at Harvard, Norman Ramsey also wanted to look for ammonia. He had a student all set to make the attempt. However, Ramsey's fellow Harvard Nobel laureate, Ed Purcell, was so sure that ammonia would not show up that he talked the student out of trying. He figured that if any did exist out there, collisions between molecules would be so rare that ammonia would be in thermal equilibrium with the cold background radiation of space and be virtually undetectable. Ed's logic was completely sound. But his basic assumption that all interstellar clouds would be of very low density, such as had been found with the 21-centimeter hydrogen line, and hence collisions too rare, turned out to be quite wrong.

For a while, I considered a first search for carbon monoxide, but its prominent spectral line at around 2.5 millimeters wavelength was too short for the relatively imprecise surfaces of the Hat Creek antennas, and amplifiers at that wavelength were not easy to get. I settled on ammonia, which has exceptionally strong lines for a given molecular concentration and prominent spectral features at about 1.3 centimeters. Even this was a rather short wavelength for radio astronomy at that time, but Hat Creek's newest antenna, 20 feet in diameter, was up to the task.

One of my Berkeley students, Al Cheung, and an excellent postdoctoral fellow, David Rank, got to work in our labs fabricating an amplifier for ammonia. Jack Welch and an engineer of the radio astronomy group, Douglas Thornton had already been making plans for work at wavelengths near the ammonia line, and they pitched in with useful advice. Together we installed the new amplifier on the 20-foot antenna at Hat Creek. Harold Weaver suggested which dust clouds were biggest and thus the best places to look. Harold's list included Sagittarius B2, a mass of dark clouds near the center of the galaxy.

In the early fall of 1968, the equipment was all ready, with an amplifier and good filters—and we looked. We first pointed the antenna in the direction of the galactic center and saw nothing. Then we went to the nearby mass of clouds known as Sagittarius B2, and there *was* an ammonia line! We tuned the amplifier and filter system to another ammonia line, and there it was, too. How easy, and how exciting! By the time we had written up our work, word had gotten out. The news media as well as scientists began buzzing us.

Finding the two ammonia lines immediately refuted the generally accepted idea that no interstellar clouds were very dense. Instead of the densities of 1 to 10 atoms per cubic centimeter, implied by measurements of the 21-centimeter hydrogen line and presumed to apply everywhere, there had to be more than 1,000 molecules per cubic centimeter in the clouds we observed. Only such molecular densities could provide enough collisions to excite ammonia to the extent implied by the radiation we measured. Such densities are still lower than the best vacuum that we have produced on Earth—but much higher densities than what was generally anticipated in space. That much gas, plus dust located in the same interstellar clouds, was adequate to shield molecules in the inner parts of the clouds from ultraviolet radiation that otherwise would tear the molecules apart.

Many experienced astronomers objected that the molecular density deduced from the ammonia spectra was so great that the cloud of gas and dust would collapse gravitationally before very long. That is, the molecules would gravitationally attract each other and all fall together to make stars. Clearly, there were many such clouds and they couldn't be collapsing as fast as theory indicated they should. So, our measurements made no sense given the ideas current at the time. To this day, just how the ammonia molecules are made, as well as the large number of uncollapsed clouds, are both still not well understood.

What molecules to look for next? There were no obvious, easy targets. Carbon monoxide and hydrogen cyanide produced wavelengths of only a few millimeters or shorter; they would take some special effort because we had no equipment already suited to work with such wavelengths. Water had a spectral line very near those of the ammonia we had found, but the excitation of water energy levels that produced this wavelength would take much higher densities and higher temperatures than seemed likely in interstellar clouds, even after the new ammonia results. However, our group—Jack Welch, Dave Rank, Al Cheung and I—thought we might as well look for water. After all, further surprises could happen. And happen they did—the water line also showed up. We found our first water in Sagittarius B2, the same place where we had first found ammonia. Al had simply calculated how to set up the equipment to look for the water line, telephoned the observer on duty at Hat Creek, and the next morning he reported it was there just like the ammonia lines, but a little stronger. Why it was there was not clear, but there it was.

Some time after we had found ammonia, Norman Ramsey ran into me at a meeting. He congratulated me on finding ammonia in space but could not restrain himself from remarking, "You know, Ed Purcell kept me from discovering ammonia before you did." And he told me the story mentioned earlier.

We were lucky we were not beaten to the discovery of water, too. We later learned that Lew Snyder and David Buhl, two young postdocs at the National Radio Astronomy Observatory in West Virginia, which had the country's biggest antenna, had shortly before that asked to use it to look for water. The committee in charge, which did not then know that ammonia was to be found, felt a search for water would be a foolish waste of antenna time and turned them down.

After water had turned up in Sagittarius B2, we of course wanted to search other sources to see if it was more widespread. Al Cheung was the graduate student doing his thesis on these spectral lines, so he had the assignment of looking for water in other sources just as Christmas was approaching. The holiday season did not deter Al. One night during Christmas week, while Frances and I had most of the research group and other friends over for drinks and a good time, Al was hard at work up in northern California at Hat Creek. The party was hitting its stride when he phoned. When I asked him how things were going, his excited reply was, "It must be raining in Orion! It has a very strong water line." He had found water, lots of it. The Orion water line was 20 times stronger than the previous one, much more than we would have expected. What Cheung had found in the Orion Nebula turned out to be a huge water maser. While he sat, happy and tired, at the control panel at Hat Creek, we all poured some champagne in our kitchen in Berkeley and toasted his success. Although we suspected the water radiation must be due to maser action, to be sure we repeated the observation of Orion and other sources as quickly as practical on the 85-foot Navy antenna at Maryland Point, just outside of Washington, D.C. Our Naval Research Laboratory collaborators were Steve Knowles and Connie Mayer. This larger antenna made the Orion water line even brighter. There were a number of other sources as well. The nebula W49, a cloudy region in which new stars were forming, had water radiation about 20 times still brighter than Orion. For mere hot water to give such intensity, it would have to be at a temperature of more than 50,000 degrees centigrade! That would be impossible, so it had to be a maser. In addition, we found that the radiation brilliance varied over a few weeks time, indicating that the source was quite small and implying temperatures as high as a billion degrees if there was not indeed a maser. It seemed to be my happy fate to encounter masers and lasers, or stimulated emission, again and again. What interesting good luck!

No space masers have been found more powerful than those based on water molecules. They might be considered the most powerful radio stations known. A single astrophysical water maser can put out much more power than the total radiation from the sun, all at a single spectral line or frequency. Some are so intense they are called megamasers—enormous in power, though perhaps not much larger in size than the solar system.

Snyder and Buhl, rebuffed from finding water by the committee at the National Radio Astronomy Observatory, got their time granted on the NRAO telescope in Green Bank, West Virginia, in a hurry after we had turned up both ammonia and water. They went out hunting for other molecules and were also very successful. The first of several that they found was formaldehyde, HCHO, done in collaboration with Ben Zuckerman and Pat Palmer, young Harvard radioastronomers who had become excited by the news of molecules. Lew Snyder commented to me that after finding formaldehyde, he thinks they would have been given antenna time to even look for the flu virus if they had asked!

An ultimate, amusing maser-inspired acronym might be attached to the formaldehyde discovery. This molecule happens to be de-excited by collisions, and it hence absorbs the 3-degree background radiation (actually 2.73 K), the ubiquitous radiation left over from the big bang. Before this, its 3 degrees was assumed to be the lowest temperature of radiation to be found in the sky. Yet the formaldehyde wavelength is, because of the absorption, still colder and perhaps the darkest part of the entire spectrum. By absorbing the 3-degree microwave background, formaldehyde fits very well Art Schawlow's joking coinage of the word dasar: *d*arkness *a*mplification by *s*timulated *a*bsorption of *r*adiation.

A number of other molecules were turning up; and before long Arno Penzias and Bob Wilson of Bell Labs had outfitted a good shortwave radio telescope in Arizona, with an amplifier for 2.5-millimeter waves, and detected carbon monoxide. This molecule is very common in interstellar clouds and has perhaps been the most intensively used in studying their densities, temperatures, and motions. Today, more than 100 different types of molecules have been found in space. They include organic molecules and highly reactive ones, such as hydrogen cyanide, alcohol, ether, methane, acetic acid (or vinegar), and molecules never before known—very long linear chains of carbon atoms ending with nitrogen. These are, in fact, just the kind of molecules that biologists expect were involved in the origin of life. By no means do all the spectral lines of molecules now known at radio frequencies come from masers, but there are more than 100 different known wavelengths at which natural masers occur, and many different molecules participate in the act.

The fact that masers exist in outer space tells us a great deal about what's going on with them—their likely local temperatures, densities, velocities, shockwaves, and boiling off of material from stars. Because masers can be very powerful and not very large, they also provide good point sources, or markers, for radio astronomy. That is, radio astronomers can measure very precisely their angular positions in the sky, as well as their velocities. A spectacular example came recently.

In 1994, a group led by Jim Moran, who had moved from MIT to the Harvard–Smithsonian Center for Astrophysics in Cambridge, turned the ten radio telescopes of the Very Long Baseline Array (the VLBA, spread from the Virgin Islands to Hawaii) on a galaxy called NGC4258, about 20 million light years away, in the direction of the Big Dipper. Earlier studies had shown that the center of this spiral-armed galaxy holds some of the most powerful water megamasers known. With the VLBA interferometer, the precise locations and the Doppler shifts (or velocities) of those masers could be mapped in detail. It revealed that the masers are embedded like glittering jewels in a flattened, spinning disk of gas, some 1,000 times wider than the solar system, in the heart of the galaxy. With the masers as markers, Jim Moran, Lincoln Greenhill, and their associates showed that the rapidly flowing gas is orbiting around a mass some 37 million times larger than the sun. This stupendous aggregation of matter is packed into a black hole, a region of collapsed matter so dense that it traps everything that enters it, and allows not even light to escape. This provides some of the most clear-cut evidence to date confirming theoretical speculations that such ultramassive black holes exist in the centers of some galaxies. Their powerful gravities are the likely engines for quasars—the most brilliant objects known—and probably also explain the powerful emanations of energy from many so-called active galaxies. There are still puzzles, however, about these black holes. Three of my students, Eric Wollman, John Lacy, and Tom Geballe, some time ago, found evidence from spectroscopic measurements of Doppler effects in ionized gas that there was a large black hole a few millions times as massive as the sun in the center of the Milky Way, which is our own galaxy. Surprisingly, it is not emanating much energy, and its presence has been questioned. Unfortunately, it is not surrounded by convenient masers for measurement, as is NGC4258. Many types of measurements have now been made, however, and increasingly good evidence has been produced that our own galaxy is also centered on a large black hole. Why it is not acting somewhat like a quasar by emitting energy as material falls into it, as does NGC4258, is poorly understood.

Just as scientists on Earth moved to study and use shorter and shorter wavelengths after the first masers were made, discovery of masers in space have proceeded down the spectrum a bit toward laser wavelengths. Measurements of the carbon dioxide (CO_2) radiation of the planet Mars, at wavelengths in the infrared region near 10 micrometers, by Berkeley students Mike Johnson, Al Betz, Ed Sutton, and Berkeley postdoc Bob McLaren, showed that CO_2 infrared radiation must be somewhat amplified in the Martian atmosphere. The CO_2 molecules evidently are excited by sunlight into a lasing condition. Mike Mumma of NASA and his associates have made still more definitive measurements on the Martian CO_2 laser action, which as mentioned earlier, played a minor role in court actions over the laser patent. The Martian maser is real, although quite weak and not spectacular.

Not long after arriving at Berkeley, I started a program in infrared spectrometry of astrophysical sources. Much of the infrared spectral region is absorbed by water vapor in Earth's lower atmosphere, so the only way to make many measurements in these wavelength ranges is from the stratosphere in an airplane (or, at much greater expense, from space). Such people as Frank Low of the University of Arizona had been making observations in infrared continuum radiation from an airplane. After I went to Berkeley, it seemed to me that one ought to be able to do spectroscopic work that way as well. As a consequence, for the last 20-odd years, our group has spent a lot of time using telescopes in aircraft that NASA has been flying from the Ames Research Center in Mountain View, a 45-minute drive from Berkeley.

Aboard these flying observatories, we have primarily examined far-infrared radiation, with wavelengths from about 1/20 to 1/5 millimeter long. Initially, we flew in a small Learjet, but starting about 1976, we made frequent use of a converted C-141 four-engined jet transport, the *Gerard P. Kuiper Airborne Observatory*, which NASA recently retired while it builds a larger flying observatory in a modified Boeing 747 aircraft.

The *Kuiper*, which added so much to our knowledge of interstellar infrared absorption and emission lines, also played a critical role in the first discovery of a laser from beyond the solar system. The first, key observations came not from the stratosphere, but from the ground. An international team led by J. Martin-Pintado looked at a hot, massive star called MWC349. In its millimeter-wave spectrum were unusually strong lines of atomic hydrogen. This, the astronomers concluded, was the first clear evidence for an atomic astrophysical maser. Until then, all astrophysical masers seen were produced by molecules, usually from energy transitions related to molecular rotation. The hydrogen-atom maser action was apparently in an accretion disk that was orbiting MWC349, a star with about 26 times the sun's mass. The setting is analogous on a smaller scale to the gigantic disk that Moran's group had observed, glittering with water masers in the NGC4258 galaxy.

The mechanism for maser action arose from hydrogen recombination—nearly all the gas in the circumstellar disk is ionized plasma, with electrons knocked off the protons and moving, separated from them. Every once in a while a proton captures an electron. The result of such captures are highly excited hydrogen atoms. Enough such atoms form to support numerous maser wavelengths as they relax toward their ground, or lowest energy, state.

While masing at millimeter wavelengths was clear, and confirmed by several groups, observations of the star at optical wavelengths showed an

Figure 15. Inside the Kuiper Astronomical Observatory, a high flying airplane for observing radiation that does not penetrate Earth's atmosphere. Charles Townes is standing and the graduate student, Sara Beck, is seated at the instrument controls during observations.

entirely normal hydrogen spectrum. In between the optical and the millimeter wavelengths is the far-infrared region, a portion of the spectrum invisible from the ground. To discover the point where masing action ceased in MWC349 would take an observatory high above the ground.

In August 1995, a group led by Vladimir Strelnitski and Howard Smith (another former Berkeley student) of the Smithsonian Institution's National Air and Space Museum flew the *Kuiper* from Hawaii to California, observing this strange star. They found a recombination line in the far-infrared, at 169-microns wavelength, clearly enhanced about 6 times higher than its thermal emission.

Still more recently, Clemens Thum of Grenoble, France, and a group of European scientists using Europe's orbiting Infrared Space Observatory (ISO) satellite found that hydrogen around the star was lasing down to wavelengths at least as short as 19 microns—and thus was discovered the most powerful presently known space laser.

With so many astrophysical masers, and at least one powerful laser from outside our solar system, one must wonder why nobody had thought about such things in space more or less as soon as stimulated emission was well understood. It is almost as puzzling as the failure to invent masers or lasers earlier. Once something is found, it's obvious. It was no secret that space has low density, and so available energy sources that can fill high-energy levels of atoms and molecules provide plenty of opportunity for energy inversions. It is also clear that if more radio astronomy had been done quite early, masers may well have been seen. Technically, this could have been done in the 1930s. Such a discovery would have led to intense speculation over the physics behind this special radiation; and that, I feel sure, would have inspired people to build masers and lasers on Earth sooner than actually occurred. Masers and lasers have been common throughout the universe for billions of years and didn't have to be invented. We just didn't look into the sky carefully enough—an unturned stone hiding masers.

Way back in the early maser days, it seemed clear that the maser's greatest value to science would come from its extraordinary sensitivity and precision. Its ability to produce nearly pure, invariant frequencies and wavelengths provides a way to keep time and make measurements to an accuracy never possible before. As the possibility of lasers was realized, it became clear that both the purity and the intensity of laser beams can be fantastic. While interesting in themselves, for science the human-made masers and lasers are primarily tools; they are means, not ends.

Astrophysical masers and lasers pack inherent excitement, but in astronomy, as in other sciences, masers and lasers have been most valuable as tools. In my own work I have put them to good use; there are some things

I could not have done without laser and maser instruments. Most recently, lasers have been vital components of projects to get high-resolution images of stars and the material immediately around them. The method, which might be thought of as training a microscope on the sky, is called stellar interferometry. The basic technique predates lasers, and was first used successfully in the early part of the twentieth century by physicist Albert Michelson.

Michelson is perhaps best known for his measurement of the speed of light and his use (with Edward Morley) of another type of interferometer, in 1887, to show that the apparent speed of light is independent of one's own velocity. The latter historic experiment provided an experimental basis for relativity.

In his 70s, Michelson came to the Mount Wilson Observatory above Pasadena. He was attracted by the 100–inch Hooker Telescope, built there by George Ellery Hale and at the time, in the 1920s, the world's largest. Michelson's goal was to measure the size of stars, something no one had been able to do for any star except our own sun. A natural target was large, relatively nearby Alpha Orionis, better known as Betelgeuse, the bright red star near the top of the Orion constellation.

Michelson blocked off most of the Hooker's mirror, leaving just two widely separated spots on its surface to reflect the light from Betelgeuse. Allowing the two light beams from the star to meet and the waves to interfere with each other, he saw alternating bright and darker bands. He then mounted a 20-foot support bar on the telescope with two mirrors on it, which could be moved apart until the alternating bands faded out. This allowed him to deduce the star's diameter. In late 1920, Michelson and his assistant, Francis Pease, had the first real measurement of the size of a star! It was no larger in angular size than the smallest dot one can make with a pencil then seen about one mile away. The angular diameter, which is what Michelson measured, is about 1/20 of a second of arc. But if astronomers are correct about its distance being about 400 light years, then its diameter is over half a billion miles. This is 600 times the diameter of our own star, the sun, and 3 times larger than the diameter of the earth's orbit around it.

Interferometry, while it may call for extreme care on the part of the experimenter, and while its principle is not as intuitively understandable as ordinary optical imaging, is not all that complicated. Here is a short explanation.

If signals from one pointlike source strike two separate mirrors and are then brought together and thus combined, interesting results become possible. Because the signals are both waves, with maxima and minima in strength, they can either reinforce each other or counteract each other, depending on the relative time of travel or relative phases of the two signals. This, in turn, depends on the difference in path length from the object to the two mirrors. A difference of only one-half the wavelength of light, or one-hundred thousandths of an inch, makes the difference between addition or cancellation of the wave crests. As an object moves a minute amount in front of the two mirrors, the arrival time of its signal at one mirror varies compared to when it hits the other, making the sum of the two signals go up and down in intensity. These are called "fringes." If the object is not strictly a point, but has parts separated by only a slight amount, the maximum fringe intensity of one part can be close to the minimum of another, and the total intensity doesn't go up and down so much because the ups and downs are averaged out. This tells the observer that the object has some real size. If the object is comparable in size to the separation between a maximum and minimum of the fringe, then these variations in intensity are blurred out, and the amount of blurring gives a measurement of size.

The farther apart the two mirrors are, the more sensitive they are to small differences of position in the object and, hence, the more detail they can pick up. With several telescopes at various positions and separations, one may construct detailed, two-dimensional maps of the signal source. The whole process requires very careful control of the mirrors, reduction of any shaking to an absolute minimum, and precise measurements. Michelson's superb experimental talents allowed him to get some results. After his initial experiments with the 100-inch telescope, he built a new, specially designed interferometer at Mount Wilson that used more widely spaced mirrors to pick up stellar signals. It included a building with a sliding roof to expose two separate telescopes.

Michelson was also busy measuring the velocity of light on Mount Wilson. He died in 1931 before his last planned measurement of light velocity, in an evacuated pipe, could be completed. He also left undone most of his planned work with the new interferometer. This task was undertaken by Francis Pease, his associate. Pease stuck with the new experiment for about 10 years, but was never able to get adequate measurements of stellar size with it.

In subsequent years, other scientists have been able to measure the sizes of some stars with techniques like those of Michelson. That has not been easy because unfortunately, the precision required to do optical interferometry is daunting. One must keep the components aligned and controlled in position to an accuracy of about one millionth of an inch. But lasers now offer some ways out of such difficulty. For one, lasers can be used to keep optical components in precise positions and alignment (interestingly, it is usually an interference phenomenon that is used to monitor these). In the late 1960s, I decided to exploit this and another technology for interferometry that could be made practical by lasers: heterodyne detection of infrared radiation.

A heterodyne detector converts signals of one wavelength to another, longer, wavelength. The idea is to mix the incoming signal with a second signal, at very nearly the same wavelength, from a local oscillator. As the incoming signal varies, it comes in and out of phase with the oscillator signal. The result is a lower frequency beat that retains much of the information from the received signal, but imbedded in a lower and more easily handled frequency. This technique, commonly used in radio reception, requires a good oscillator. At wavelengths much shorter than radio waves, lasers now provide the needed oscillators.

The plan was to combine a laser oscillator's output with infrared radiation from stars, converting the stellar signal from infrared into a longer, radio wavelength. The advantage is that with a longer wavelength, one does not have to control as many distances with such high precision as when the raw infrared signal is used.

Two Berkeley graduate students, Michael Johnson and Albert Betz, went to work on the project, which even with the help of lasers and modern infrared detectors was by no means easy. I have always felt lucky to have worked with so many outstanding students. These two were examples, and they pushed the project through. A complete system, without telescopes, was put together in our Berkeley laboratory. It used carbon dioxide lasers, operating at wavelengths near 10 micrometers, as local oscillators. We started work around 1973 at the Kitt Peak National Observatory in Arizona, using a couple of little-used solar telescopes about 18 feet apart to receive the infrared radiation. We hauled the equipment from Berkeley to Kitt Peak in a truck. Once it was there, we flew back and forth from San Francisco to Tucson, the nearest big town to Kitt Peak. In addition to interferometry, the heterodyne detection system lent itself nicely also to doing spectroscopy on astronomical objects. With it, Betz and Johnson discovered the natural CO_2 laser amplification on Mars, which was mentioned earlier. They also obtained very accurate measurements of wind velocities on Mars and Venus.

After much preparation, trial, and error, our first test of the interferometer was an effort to simply detect the edge of the planet Mercury. It worked! And we celebrated with a fancy dinner together in San Francisco, which I had promised as a recognition of success.

We next turned to Betelgeuse, to the well-known variable star Mira, and to IRC+10216, a star in the constellation Leo that is one of the brightest infrared stars in the Northern Hemisphere. Another graduate student, Ed Sutton, came into the project as Mike and Al were about to get their Ph.D.

degrees. The interferometer gave us measurements of the size and some of the characteristics of the shells of warm dust around the stars, dust that had condensed from gas that the stars had previously emitted.

The Kitt Peak work was encouraging, but it was just a start—a prototype system—and it was clear we needed better equipment. For one thing, I wanted movable telescopes so that the interferometer could be optimized for sources of different sizes. Our initial applications for grants to build such telescopes were turned down by the National Science Foundation and by the Defense Department. Astronomers on the review committees did not think that infrared interferometry held enough promise to warrant spending the money required. After the disappointments experience by Francis Pease, Michelson's associate, other hard work at optical interferometry had achieved some success but had then been abandoned.

Finally, in late 1982, we received a grant of about $2.5 million from the Pentagon's Advanced Research Projects Agency to start construction of the interferometer. The Navy, feeling that interferometry might contribute to its traditional interest in measurement of stellar positions (a legacy of the days when ships depended on stellar navigation), also provided solid support and help. The by-then experienced scientist Ed Sutton and I began work building the equipment at Berkeley's Space Sciences Laboratory, on the hill east of campus. We were soon joined by young postdocs Bill Danchi from Harvard and Manfred Bester from Cologne, Germany.

Each of the interferometer's two telescopes has a 65-inch mirror and is mounted in a trailer that can be moved by truck. The telescope design is rather unusual. Each is fixed in a horizontal position. Pointing is done by a big, flat, steerable mirror that is 85 inches across. Each flat sits in the open, in the middle of its trailer. It reflects the signal to a 65-inch parabolic mirror, which focuses the infrared light and sends it back, through a hole in the flat, to the detection system.

The ideal place for optical or infrared interferometry is northern Chile. The mountains there have the best seeing, or the least turbulent atmosphere, that astronomers have found. Good seeing is essential. Blurriness introduced by atmospheric distortion both prevents all the light from being focused well and also varies the phase of the light waves, often ruining any chance to get high-quality interference.

In the continental United States, the best seeing we know is right where Michelson worked, on Mount Wilson. So, in 1988, we towed our two telescopes down Interstate 5, up the Angeles Crest Highway to the mountain top, and set them up as the Infrared Stellar Interferometer, next to Michelson and Pease's old building. In the years since, we have occasionally cannibalized that deteriorating old structure when we needed pieces of steel or wood for our project.

Figure 16. The christening at the University of California of our first large movable telescope on a trailer, one unit of the Infrared Spatial Interferometer, which maps the details of stellar shapes and the clouds around stars. In operation, laser beams shine back and forth between the two mirrors. Left to right are Charles Townes, electronics technician Walter Fitelson, and physicists Edmund Sutton, William Danchi, and Manfred Bester.

We move the telescopes around among a series of concrete pads to vary the distances and orientation of the line between the two telescopes. The astronomical community is beginning to be convinced that interferometry is now promising. Recently, we received from NSF funds to build a third telescope, which will increase markedly the amount and quality of interferometry. We will then have three pairs of telescope eyes with which to look instead of only one pair. We may tear down remains of the old Michelson–Pease building and pour a concrete pad for our telescopes on the spot. If so, we will have to call it Michelson Station and affix a plaque or other marker noting its history.

So far, our infrared interferometer has been used to measure the size of some stars. Perhaps more important, infrared wavelengths reveal the behavior of much cooler material immediately surrounding stars and flowing out of them. It has been known for some time that many stars, particularly older ones, emit quantities of gas and are surrounded by dust shells. This is true of Mira, the most famous of stars that vary regularly in brightness, and also of Betelgeuse and many others. With our infrared interferometer, we have found that much of the material flowing from these stars is not emitted continuously, but comes out episodically. It is as though huge, random events blast material off these stars and out into space. Betelgeuse, for example, has little dust very close to it. Nothing much, it appears, has flowed out of the star for some time. But about ten stellar diameters away is an expanding shell of dust and gas. Measurements indicate that this shell was probably produced about 1943. Another more distant shell may have been produced in 1836. In that year the great English astronomer, John Frederick Herschel, noticed and called everyone's attention to changes in the brightness of Betelgeuse that were so remarkable they could easily be seen by eye. Other stars, such as Mira, have dust clouds only a couple of stellar diameters away. Fortunately, we were watching Betelgeuse in 1994, during a crucial period when it produced another outburst of gas and dust. It was not as big as that seen by Herschel, but plenty big enough to be detected with our interferometer.

The interferometer has led to yet another encounter with natural masers and their physics. As discussed earlier, for some time, various groups have observed masers that were due to water (H_2O), OH, or silicon monoxide (SiO), which are associated with stars. Just where and why they occur has not been so clear. What we find with our interferometer is that water and OH masers are most common around stars that have gas and dust quite close by, where energy and density can be high. Those where the gas and dust clouds are farther away do not generally produce such masers. We now know that SiO masers are produced still closer to the stars, in the very hot atmosphere immediately surrounding them. A careful measurement of the star VX Sagittarii, a well-known star that is bright at infrared wavelengths, was made by Bill Danchi and others in our research group. Combined with measurement of the SiO masers by Jim Moran and Lincoln Greenhill of Harvard, who used a radiofrequency interferometer, those results showed that the SiO masers were much closer to the star than any dust or molecular cloud, so they must occur in the outer stellar atmosphere itself.

And my favorite molecules—ammonia, along with others—have roles to play in infrared interferometry. A graduate student, John Monnier, is developing an interferometry study of the spectra of ammonia molecules in the clouds surrounding old stars to see just where, and perhaps how, the ammonia forms in material flowing from such stars. Possibly this will help in the still-standing puzzle of how ammonia is formed and why we find it in interstellar clouds.

Interferometry is still essentially a young field. A number of other groups have formed to pursue it, particularly using visible or short-wavelength infrared radiation and without heterodyne-type detection. These now include a group in Australia, one at an observatory in the hills above

Nice, France, and others at the Jet Propulsion Laboratory and at Caltech in Pasadena, the Naval Observatory, Harvard, and Georgia State University. Such interferometry is difficult, but lasers have helped make them practical, and those of us who work on interferometry believe that modern technology combined with clever and hard work will make it very rewarding. We all hope to see more and more of the detailed behavior of stars and other heavenly objects. There is much more to be discovered.

10

GLANCES BOTH BACKWARD AND FORWARD

The biochemist Albert Szent-Gyorgi has said "Discovery is seeing what everybody else has seen, and thinking what nobody else has thought." The laser discovery seems to fit this image; it is built upon ideas that were long known, at least to some. Yet the ideas had to be assembled in a novel way and the value of doing that had to be recognized.

I see discovery as still more multifaceted than Szent-Gyorgi's perceptive description. We explore. What path to explore is important, as well as what we notice along the path. And there are always unturned stones along even well-trod paths. Discovery awaits those who spot and take the trouble to turn those stones. In that case, one may see what no one has seen before. And some aspects of maser and laser discoveries, such as the remarkably pure frequencies produced, or the enormous concentration of power and its effects, also fit this image. We had to undertake making masers and lasers before these phenomena were clearly seen.

If Gordon, Zeiger, and I had not followed the path we did and made a maser, and if Schawlow and I had not gotten together, how long would development of the maser and laser have been delayed? My guess is not more than a decade, or possibly two. After all, within a little more than a decade after the first man-made maser, some of those in space became known through a quite different route, radio astronomy. The radio astronomy path would have allowed us to discover masers, which have been in the heavens for billions of years, and probably would have led on to lasers. Nevertheless, discovery of the laser one or two decades sooner rather than later by this possible route has made a big difference in the fast-moving science and technology characteristic of the world today.

In many cases, it is not just new insight, but new tools or technology that open up more penetrating exploration than was previously possible—a better microscope, or telescope, or measuring device, or material. Our present scientific knowledge and technology make possible the next steps toward new science and technology, which in turn lead us on still farther. The development of quantum electronics has provided remarkable new tools and repeatedly demonstrated this process.

Individuals sometimes ask me whether a story such as that of the laser can happen in science today. They raise the question partly because there's an impression that at least the so-called physical sciences are almost finished—"we know all the basic science"—and partly because much of science appears to come nowadays from large teams of people—with large and expensive equipment. Is there still room for individual discovery of importance? My answer is that every scientific discovery is different in detail, and essentially unpredictable, but that there will be many more. There is much that we don't understand; in many cases we don't understand that we don't. And the really surprising discoveries will probably depend primarily on individuals, not teams or committees even though the individual may be part of a team. As I was finishing my graduate training, the field of optics was thought to be essentially understood and finished—no longer exciting for researchers. A little more than 20 years later, the laser gave it a rebirth, with both new science such as nonlinear optics, and important new technology emerging for both science and industry.

Among the many things we don't presently understand, there are such questions as how our universe really began, why the physical constants have the values they do, and how to make gravitational and quantum theory consistent. We hardly even understand how to go about trying to find out. New phenomena are popping up in solid, gaseous, and liquid materials, in astronomy, in many fields. Biology beckons with manifold interesting puzzles. Some clearly recognizable paths toward discovery can be defined, but they may be less fruitful than energetic, persistent, and flexible curiosity—simply exploring and trying to understand. Much that we haven't yet imagined, and thinking we have not yet thought, remains in both science and technology.

I myself feel very fortunate to be able to spend my life exploring and to be a part of the scientific community, enjoying science and the intimate, powerful connections that it turns up. Scientific principles are so general and pervasive that they continually show up as familiar friends in new territory—or with exploration in any direction. I am both thrilled and intrigued by nature's beauty. Somehow, essentially every aspect of nature can be inspiring and beautiful. A calm sea and a stormy sea are both strikingly esthetic and stimulating. So is the structure of an atom, a field fresh with flowers, a desert, an insect, bird, fish, star, galaxy, or the mysteries of a black hole. As I have had a chance to explore and try to understand, I feel enriched—not just by the usefulness of science, but by its awesomeness, connectedness, and the beauty of all its dimensions. Scientific exploration is indeed fun, and thinking over the experiences or the pathways that my colleagues and I have excitedly enjoyed is an occasion to be thankful.

INDEX